Simon Newcomb

The elements of the four inner planets and the fundamental

constants of astronomy

Simon Newcomb

The elements of the four inner planets and the fundamental constants of astronomy

ISBN/EAN: 9783337276584

Printed in Europe, USA, Canada, Australia, Japan

Cover: Foto ©berggeist007 / pixelio.de

More available books at **www.hansebooks.com**

THE ELEMENTS

OF THE

FOUR INNER PLANETS

AND THE

FUNDAMENTAL CONSTANTS OF ASTRONOMY

BY

SIMON NEWCOMB

Supplement to the American Ephemeris and Nautical
Almanac for 1897

WASHINGTON
GOVERNMENT PRINTING OFFICE
1895

PREFACE.

THE diversity in the adopted values of the elements and constants of astronomy is productive of inconvenience to all who are engaged in investigations based upon these quantities, and injurious to the precision and symmetry of much of our astronomical work. If any cases exist in which uniform and consistent values of all these quantities are embodied in an extended series of astronomical results, whether in the form of ephemerides or results of observations, they are the exception rather than the rule. The longer this diversity continues the greater the difficulties which astronomers of the future will meet in utilizing the work of our time.

On taking charge of the work of preparing the *American Ephemeris* in 1877 the writer was so strongly impressed with the inconvenience arising from this source that he deemed it advisable to devote all the force which he could spare to the work of deriving improved values of the fundamental elements and embodying them in new tables of the celestial motions. It was expected that the work could all be done in ten years. But a number of circumstances, not necessary to describe at present, prevented the fulfillment of this hope. Only now is the work complete so far as regards the fundamental constants and the elements of the planets from Mercury to Jupiter inclusive. The construction of tables of the four inner planets is now in progress, those of Jupiter and Saturn having already been completed by Mr. HILL. All these tables will be published as soon as possible, and the investigations on which they are based are intended, so far as it is practicable to condense them, to appear in subsequent volumes of the *Astronomical Papers of the American Ephemeris*. As it will take several years to bring out these volumes, it has been deemed advisable to publish in advance the present brief summary of the work.

The author feels that critical examination of this monograph may show in many points a want of consistency and continuity. The ground covered is so extensive, the material so diverse as well as voluminous, and the relations to be investigated so numerous, that no conclusion could be reached on one point which was not liable to be modified by subsequent decisions upon other points. The author trusts that the difficulties growing out of these features of the work, as well as those incident to the administration of an office not especially organized for the work, will afford a sufficient apology for any defects that may be noticed.

NAUTICAL ALMANAC OFFICE,

U. S. Naval Observatory, January 7, 1895.

CONTENTS.

v

Page.

CHAPTER IV.—COMBINATION OF THE PRECEDING RESULTS TO OBTAIN THE MOST PROBABLE VALUES OF THE ELEMENTS AND OF THEIR SECULAR VARIATIONS FROM OBSERVATIONS ALONE.

CHAPTER V.—MASSES OF THE PLANETS DERIVED BY METHODS INDEPENDENT OF THE SECULAR VARIATIONS, WITH THE RESULTING COMPUTED SECULAR VARIATIONS.

ELEMENTS AND CONSTANTS.

CHAPTER 1.

GENERAL OUTLINE OF THE WORK OF COMPARING THE OBSERVATIONS WITH THEORY.

1. In logical order, the first step in the work consists in the reduction of observed positions of the Sun and planets to a uniform equinox and system of declinations.

The adopted standard of Right Ascensions was that originally worked out in my paper on the Right Ascensions of the fundamental stars, found in an appendix to the *Washington Observations for 1870*, and extended to a fundamental system of time stars in the catalogue published in Vol. 1 of the *Astronomical Papers of the American Ephemeris*. This system coincides closely with that of the *Astronomische Gesellschaft* and the *Berliner Jahrbuch*, about the epoch 1870, but the centennial proper motion is greater by about $0^s.08$.

In Declinations, the adopted standard was that of Boss, which has been used in the American Ephemeris since 1881, and on which is based the catalogue of zodiacal stars just referred to. But as Declinations generally are not immediately referred to fundamental stars, the method of reducing observations to this system in Declination was not entirely uniform.

Observations used.

2. The following is a general statement of the observations used, and the extent to which they were corrected, or re-reduced.

Greenwich.—Dr. AUWERS courteously supplied me with the results of his re-reduction of BRADLEY's observations both of the Sun and planets. From the beginning of MASKYLENE's work until 1835, the Greenwich observations were completely re-reduced, utilizing, so far as possible, AIRY's reductions. The data necessary for these observations were discussed in Prof. SAFFORD'S paper, Vol. II, pt. II, which paper was prepared for this purpose. In the case of the Greenwich observations from 1835 onward, it was deemed sufficient to apply constant corrections to the Right Ascensions, determined from time to time by comparisons of the adopted Right Ascensions with the standard ones. In the case of the Declinations, Boss's special tables were used, but in the later years it was judged sufficient to apply the constant correction necessary for reduction to Boss's standard.

Palermo.—PIAZZI'S observations of the Sun and Planets were completely re-reduced, the zero point of his instrument being determined from the observed Declinations.

Paris.—LEVERRIER'S reduction of the Paris observations from 1801 onward was made use of, applying the correction necessary to reduce the results to the adopted standard.

Königsberg.—BESSEL'S clock corrections were individually corrected by the new positions of the fundamental stars, so that practically the Right Ascensions may be considered as completely re-reduced.

In the case of the other observatories, it was deemed sufficient to determine, by a comparison of the adopted or of the concluded Right Ascensions and Declinations of the fundamental stars with the standard catalogue, what common corrections were necessary for reduction to the standard. When, however, the period was covered by Boss's tables, the correction which he gives as varying with the Declination was applied. After more mature consideration, I am inclined to think it would have been better to apply a constant correction to the Declinations in every case, except those where the change with the Declination was quite large.

Although these processes were somewhat heterogeneous, it is believed that the main object of referring the Declinations to a system of which the error would be a uniformly varying quantity was fairly well attained. The subsequent determination of this error both in Right Ascension and Declination is a necessary part of the work.

The following is a list of the observatories whose observations of the Sun and Planets were included in the work:

Greenwich _____ 1750–1892
Palermo _____ 1791–1813
Paris _____ 1801–1889
Königsberg _____ 1814–1845
Dorpat _____ 1823–1838
Cambridge _____ 1828–1844
Berlin _____ 1838–1842
Oxford, Radcliffe _____ 1840–1887
Pulkowa _____ 1842–1875
Washington _____ 1846–1891
Leiden _____ 1863–1871
Strassburg _____ 1884–1887
Cape of Good Hope _____ 1884–1890

The number of the meridian observations of the Sun, and
of the planets Mercury, Venus, and Mars, actually included in
the work is approximately as follows:

The Sun _____ 40, 176
Mercury _____ 5, 421
Venus _____ 12, 319
Mars _____ 4, 114

Total _____ 62, 030

Semidiameters of Mercury and Venus.

3. The reduction of the semidiameter of the planets was a
point to which special attention was given. In the case of
Mercury, the adopted semidiameter at distance unity was $3''.34$.
The values adopted by the various observatories in reducing
their observations varied so little from this that in cases where
the original reductions were accepted no correction was applied
for the difference. So, also, when the observers applied a correction for reducing the observed center of light to the actual
center of the planet, no revision of this reduction was made.
Such was supposed to be the case with the Paris observations.
When the published Right Ascension was that of the center
of light simply, a reduction to the true center was computed
by the empirical formula used in the Washington observations.
If we put i for the angle between the Earth and Sun as seen
from the planet, then $1 + \cos i$ will represent the fraction of

the apparent transverse diameter of the planet that is illuminated by the Sun. It was assumed that when the illumination was such that the thickness of the crescent approached zero, the point observed would be two-thirds of the way from the center of the planet to the limb, and that when the planet was dichotomized the center of observation would be five-twelfths of the way from the center to the limb. These conditions, with the added one that when the planet was fully illuminated the correction should vanish, suggested the employment of the formula

$$\text{Correction} = \text{semidiameter} \times \frac{(1-\cos i)(5+\cos i)}{12}$$

This correction was to be multiplied by the sine or cosine of the angle which the line of cusps made with the meridian to reduce it to Right Ascension and Declination respectively.

The correction being practically the same whenever the Earth and planet return to the same positions in anomaly, it is possible to embody it in a table of two arguments, one depending on the longitude of the Earth, the other on that of the planet. Actually, however, the table was arranged in a more convenient form, in which one argument is the date at which Mercury last passed perihelion, and the other, its mean anomaly. Owing to the importance which this correction may assume, a partial transcript of the table actually employed for the reduction in Right Ascension is given on the next page. Read horizontally, the numbers show the corrections of the argument through one revolution of the planet. Vertically, they may be regarded as giving the successive corrections corresponding to any one position of the planet, while the Earth goes through a complete revolution. The table as actually used extended to every 10°, but the values for every 60° of mean anomaly will suffice to show the general magnitude of the correction.

The correction to the Declination was embodied in a similar table, which it is not deemed necessary to print at present.

In the case of Venus, it seems scarcely possible to decide upon a value of the semidiameter, or a law of its apparent change, which should apply to all parts of the orbit. After a careful examination of the data, it was decided to reduce all the observations with the semidiameter

$$\frac{8''.75}{\triangle} + 0''.20$$

when made with modern instruments, and to use a value $0''.3$ greater in earlier observations. The actual reductions of all

Correction for defective illumination of Mercury in R. A. Arguments: Date of perihelion passage at side, and mean anomaly "g" at top.

$g=$		0°	60°	120°	180°	240°	300°	360°
		s	s	s	s	s	s	s
Jan.	0 --	+.19	—.16	—.07	—.03	—.01	.00	+.03
	10 --	.16	—.18	—.09	—.04	—.01	.00	.02
	20 --	.14	—.21	—.11	—.05	—.02	.00	.02
	30 --	.12	—.19	—.13	—.06	—.03	.00	+.01
Feb.	9 --	.10	—.17	—.15	—.08	—.04	—.01	.00
	19 --	.08	--------	—.18	—.10	—.05	—.01	.00
Mar.	1 --	.06	+.16	—.21	—.12	—.06	—.02	.00
	11 --	.05	.16	—.24	—.15	—.08	—.03	.00
	21 --	.04	.15	—.26	—.18	—.10	—.04	.00
	31 --	.03	.14	--------	—.20	—.12	—.06	—.01
Apr.	10 --	.02	.12	+.23	—.22	—.15	—.07	—.01
	20 --	.02	.10	.20	--------	—.18	—.09	—.01
	30 --	+.01	.08	.18	+.24	—.21	—.11	—.02
May	10 --	.00	.06	.15	.22	—.17	—.13	—.03
	20 --	.00	.05	.12	.20	—.12	—.16	—.04
	30 --	.00	.04	.10	.17	--------	—.18	—.05
June	9 --	.00	.03	.09	.14	+.18	—.20	—.06
	19 --	.00	.02	.07	.12	.16	—.20	—.07
	29 --	—.01	.01	.05	.09	.15	—.20	—.09
July	9 --	—.01	.01	.04	.07	.13	--------	—.11
	19 --	—.01	+.01	.03	.05	.11	+.16	—.12
	29 --	—.02	.00	.02	.04	.09	.16	—.14
Aug.	8 --	—.03	.00	.01	.03	.07	.16	—.16
	18 --	—.04	.00	.01	.03	.06	.14	—.18
	28 --	—.05	.00	+.01	.02	.05	.13	--------
Sept.	7 --	—.06	.00	.00	.02	.04	.11	--------
	17 --	—.07	—.01	.00	+.01	.02	.09	--------
	27 --	—.09	—.01	.00	.00	.02	.07	+.20
Oct.	7 --	—.11	—.02	.00	.00	+.01	.05	.18
	17 --	—.12	—.02	.00	.00	.00	.04	.16
	27 --	—.14	—.03	—.01	.00	.00	.03	.13
Nov.	6 --	—.16	—.04	—.01	.00	.00	.02	.11
	16 --	—.18	—.06	—.01	.00	.00	+.01	.09
	26 --	--------	—.08	—.02	.00	.00	.00	.07
Dec	6 --	--------	—.10	—.03	—.01	.00	.00	.06
	16 --		—.12	—.05	—.01	—.01	.00	.05
	26 --	+.20	—.15	—.06	—.02	—.01	.00	.04
Jan.	5 --	+.18	—.17	—.08	—.03	—.01	.00	+.03

the principal series of observations were corrected to this value of the element in question.

Observations of the estimated center of Venus, when made more than one hundred days from superior conjunction, were rejected altogether; when made within that limit, the point observed was assumed to be the center of gravity of the illuminated portion of the disk, considered as a plane figure, and the necessary reduction to the center was always applied.

A similar correction was applied to observations of the estimated center of Mars. The Paris results, after 1830, and the later Greenwich and Washington results, are published with the reduction for center of light already applied, and in these cases the published corrections were not changed.

Tabular places.

4. The tabular elements of the planets adopted for correction were those of LEVERRIER's tables. These tables having been continuously used in Astronomical Ephemerides since 1864, it was judged more convenient to adopt the theory on which they were based as the provisional one to be corrected than it was to construct a new provisional theory. As the tables in their original form are extremely cumbrous to use, the theory was partially reconstructed by making manuscript tables of the principal perturbations, which were, however, carried only to tenths of seconds. With these tables the places of the planets were computed for dates previous to 1864.

As places of the Sun were necessary not only for direct comparison with observations of the Sun, but also for the geocentric places of the planets, an ephemeris of the Sun's longitude and radius vector was prepared for the entire period 1750–1864 to every fifth day, the lunar perturbation being omitted and afterward applied for each date when required.

The method of deriving the final tabular places varied with circumstances. When there was no accurate ephemeris available for comparison, which was the case before 1830, it was necessary to compute a completely independent set of tabular geocentric places. Sometimes these places were computed for the moment of the individual observations, but more generally, when the observations occurred in groups, an ephemeris was

computed in order that the work might be checked by differences. After 1830 it was common to compute an ephemeris
for intervals of three, five, or ten days, thus deriving the corrections necessary to reduce the published ephemerides of the
Berliner Jahrbuch or of the *Nautical Almanac* to those derived
from LEVERRIER'S tables.

Until this plan was mapped out, and work well in progress
upon it, it was not noticed that the planetary masses adopted in
LEVERRIER'S tables were so diverse that corrections to reduce
the geocentric places to a uniform system of masses would be
necessary. Although theoretically the necessary reductions
were very simple, I can not but feel that the application of
such corrections involves more or less doubt and uncertainty,
and that it would have been better to have constructed provisional tables based on uniform masses quite independent of
those of LEVERRIER.

In *Annales de l'Observatoire de Paris*, Vol. II, LEVERRIER
gives the following values of the masses used by him as the
basis of his provisional theory:

$$\text{Mercury} \dots \frac{1}{3\,000\,000} = .000\,000\,333\,.\,.$$

$$\text{Venus} \dots \frac{1}{401\,847} = .000\,002\,4885$$

$$\text{Earth} \dots \frac{1}{354\,936} = .000\,002\,8174$$

$$\text{Mars} \dots \frac{1}{2\,680\,337} = .000\,000\,373\,087$$

The following table shows the factors by which these masses
were multiplied in the cases of the several planets in LEVER-
RIER'S final tables. They were controlled by induction from
the numbers of the tables themselves, the result of which was
found in all cases to agree with the statements in the introduction to the tables.

In the last line of the table is shown the factor used in the
present provisional theory.

	Mercury.	Venus.	Earth.	Mars.
In tables of—				
The Sun	I	1. 004		0. 895
Mercury		I	I	
Venus	I		I	I
Mars		0. 975	1. 0026	
Present work	I	I	I	0. 8657

As in the actual work the masses of Mercury and Venus were to be determined from the observed periodic perturbations which they produced, it was necessary that the perturbations produced by them should all be carefully reduced to the adopted standard. The reduction was less necessary in the case of Mars, but was carried through all the work relating to the Sun.

Comparison of observations and tables.

5. The result of each separate observation of each body was compared with the tabular result thus derived. The residuals were then taken and divided into groups. The interval between the extreme dates of each group was always taken so short that it could be presumed that the mean of all the residuals would be the correction for the mean of all the dates. The general rule was that the interval should not exceed four or five days in the case of Mercury, or six or eight days in that of Venus, and that not more than six or eight observations should be included in a single group. In taking these means, weights were assigned to the results of each observatory founded on the discordance of its residuals. Then to each mean a weight was again assigned equal to the sum of the weights of the individual residuals when these were few in number, but not allowed to exceed a certain limit, how great soever might be the sum of the individual weights.

Equations of condition.

6. Each mean result thus derived formed the absolute term of an equation of condition for correcting the tabular elements. The number of these equations was as follows:

	Equations.
The Sun	11, 676
Mercury	3, 929
Venus	4, 849
Mars	1, 597

In forming the equations of condition from observations of the planets, I adopted the system suggested in the introduction to Vol. I of these publications, namely, the determination of the solar elements not only from observations of the Sun itself, but from observations of each of the planets. The reason for this course is quite simple and obvious. An observation of the position of a planet as seen from the Earth is the exact equivalent of an observation of the Earth as seen from a planet, and thus depends equally upon the elements of both orbits. Hence, whatever elements of the Earth's orbit could be determined by observations made from a planet can equally be determined by observations made upon the planet. A strong reason for proceeding upon this plan was found in the very large errors, both accidental and systematic, to which observations of the Sun are liable.

The advantages, however, have not proved relatively so great as were anticipated. The eccentricity and perihelion of the Earth's orbit come out in the solution of the normal equations as functions of those of the planetary orbit to so great an extent that their weight is much less than that which would correspond to independent determinations from the same number of observations. On the other hand, the determination of these elements from observations of the Sun proved to be much more consistent than was expected, thus indicating a high degree of precision.

The case is different with the Sun's mean longitude referred to the Stars. Here systematic and personal errors enter so largely that the results from Mercury and Venus appear to be rather more reliable than those from the Sun itself. In the case of these planets it fortunately happens that the weight of the result derived for the Sun's mean longitude is not materially diminished by the uncertainty of the corresponding element of the planet, the errors of the two mean longitudes being nearly separated in a series of observations equally distributed around the orbit.

The systematic errors in observations of the Sun rendered it unadvisable to determine the elements of the Earth's orbit from observations of the Sun by a single system of equations. The solar observations, therefore, were classified according to the observatory where made, and divided into periods rarely exceeding eight years in length. The elements are separately derived from the observations of each period. This system has the advantage of eliminating to a large extent the injurious effect of systematic and personal error upon the eccentricity and perihelion of the Earth's orbit, and also enabling us to judge of the precision of the corrections to those elements by the discordance among separate results.

Meridian observations of the Sun and Planets are referred to the fundamental stars, while the Right Ascensions of the latter are referred to the equinox, the position of which has heretofore depended on observations of the Sun. The adopted position of the fundamental stars therefore comes in, to a certain extent, as the basis of the work, and the constant parts of their systematic corrections are among the results to be derived.

Thus, in the case of the equations pertaining to the three planets, the following corrections were introduced as unknown quantities:

Correction of the mass of Mercury or of Venus.

Corrections to the elements of the orbit of the planet observed.

Correction of the obliquity of the ecliptic.

Corrections to the Sun's mean longitude, eccentricity, and longitude of perihelion.

Common corrections to the adopted Right Ascensions and Declinations of the fundamental stars.

In the case of Mercury an adopted hypothetical correction of the ratio of the radius vector of the planet to that of the Earth was also included in the equations, although little doubt could be felt that the true value of such a quantity must be zero. The reason for introducing it will be explained hereafter.

Determinations of the masses and secular variations.

7. The secular variation of all the preceding elements, the mean distances excepted, was also introduced into the equations from observations of the planets. In addition to the above elements, the mass of Venus appeared in the equations

derived from observations of the Sun, Mercury, and Mars, and
the mass of Mercury in the equations derived from observations of Venus. The coefficients of the masses, however,
depended wholly upon the periodic perturbations.

Were it quite certain that the secular variations arise
wholly from the masses of the known planets, the masses
could of course be derived from these variations, and the latter would appear in the equations of condition only through
the mass itself. On this hypothesis the secular variations
would not appear in the equations, but only the masses. But
it is well known that the perihelion of Mercury is subject to a
secular variation which can not be accounted for by any admissible masses of the known disturbing planets. The same
thing may well be true of the secular variations of the other
elements. It is therefore necessary, in the absence of a known
cause for such deviations, to derive the masses of the planets
independently of the secular variations. In the case of Mars
the mass is obtained with all necessary precision from the satellites. It is, however, different in the case of Mercury and
Venus. Here no resource is left us but to determine them
from the periodic inequalities. As the inequality produced by
Venus in the Earth's longitude is rarely more than eight seconds, it might seem that the coefficient would be too small to
obtain a sufficiently precise value of the mass. But in the
case of observations upon the Sun, Mercury, and Mars the
error of the determination of the mass in question may be
almost indefinitely reduced by multiplication and extension
of the observations without danger of systematic error.

To illustrate this, let us suppose the Sun's longitude to be
determined with a meridian instrument only once a year, say
at equal intervals of three hundred and sixty-five days. Let
the longitudes thus observed be compared with an ephemeris
in which the elements are affected with only slight errors.
Leaving out of consideration the periodic perturbations produced by the planets, the comparison of the observed longitudes with the tabular ones through an entire century should
be nearly constant. Any error affecting all the longitudes
alike would appear as a constant. The errors of mean motion

would vary uniformly with the time. Thus the other elements would be nearly constant, and could be still more approximately represented by a slight apparent secular variation.

Now let the disturbing action of a planet, say Venus, be introduced. We should then have a series of deviations from the law of uniform increase, which would enable us to evaluate the mass of the planet. The value of this mass thus derived would not be affected by any systematic error common to all the observations, nor even by such an error which varied uniformly with the time. Nor would small errors in the adopted elements of the Sun have any effect upon the result.

If this would be the case for observations made only at a certain point of the orbit, *a fortiori* would it be the case for the observations made at various points of the orbit, since any tendency to a systematic effect of the errors of observation would thereby be ultimately eliminated.

Considerations almost identical apply to the case of observations upon either of the planets when we consider the action of the other planet upon the planet observed and upon the earth. But they do not apply to the case of the action of the earth itself upon the observed planet, or *vice versa*. For example, in the case of observations of Venus, we may suppose that all observations made when Venus is at a certain point of its relative orbit, near inferior conjunction, say one month before inferior conjunction, are affected with a certain error common to all observations made at that point of the orbit. Since the perturbations produced by the third planet will in the long run have all values, positive and negative, for these several observations, the systematic error in question will not affect the ultimate value of its mass. But the perturbations of Venus produced by the Earth, as well as those of the Earth produced by Venus, will not have all values in such a case, but only special ones dependent on the relative position. Hence, determinations of these masses might be affected by errors of the kind in question. We conclude, therefore, that the mass of the Earth can not be satisfactorily determined by the periodic perturbations which it produces in the motion of any planet, nor that of Venus by observations on Venus through its periodic perturbations of the Earth.

In the solution of the equations of condition the method of least squares has been used throughout, the arrangement of the work, the choice of quantities to be corrected, and the accuracy of the coefficients being so chosen as to minimize the great mechanical labor of making the necessary multiplications. The adoption of this method was necessary in order to separate, so far as possible, the various unknown quantities and show to what extent their values were interdependent. By no other method of combination could so large a number of unknown quantities have been separately determined in a way which would have been at all satisfactory. On the other hand, in combining the final results and deciding upon the values of the corrections to be adopted, the method has not always been applied, for reasons which will be developed in Chapter IV.

Introduction of results of observations on transits of Venus and Mercury.

8. In the case of Mercury and Venus the observed transits over the Sun give relations between the corrections to the elements more accurate than those ordinarily derivable from meridian observations. This is especially the case with Venus. The value of these observations is greatly increased by the fact that they are made when the planet is near inferior conjunction, and therefore nearest to the Earth, and in a point of the relative orbit where meridian observations are necessarily most uncertain. In the case of Venus the error of the heliocentric place will be more than doubled in the case of the geocentric place during a transit. As, however, the observation of a transit gives no one element, but only an equation of condition between the values of all the elements at the epoch, the only way of treating it is to introduce the result as such an equation, with its appropriate weight. The determination of the proper weight is a difficult matter. The systematic errors of meridian observations are such that the theoretical value of the weights assignable to so great a mass as we have discussed would be entirely illusory. In fact so great is the weight assignable to the observed transits of Venus that if we should regard the results of each transit as a condition to

be absolutely satisfied we should not be dangerously in error. I conclude, therefore, that there is more danger of assigning too small than too great a weight to these observations.

In order to determine what change was produced in the results by the use of the observed transits over the sun's disk, two separate solutions of the equations of condition for Mercury and Venus were made. In the one, termed solution A, the meridian observations alone were used; in the other, termed solution B, the combined equations formed by adding the normal equations derived from the transits to those given by the meridian observations were used.

In the case of solution A it was originally supposed that by using the mean epoch of all the observing in the case of each planet as that from which the time was to be reckoned, the normal equations for the secular variations would be almost completely separated from those for the corrections to the elements themselves. The separation would be complete were the observations at different epochs similarly distributed around the orbit. But, as a matter of fact, it was found that the accidental deviations from this symmetry were so considerable that the separation could not be regarded as complete. The solution was therefore made by successive approximations, the terms depending on the secular variations being in the first approximation dropped from the normal equations for the corrections to the elements, and afterwards included when approximately determined, and *vice versa*.

In the case of solution B, in which the transits were included, such a separation did not occur, and the equations were solved in the usual rigorous way for all the unknown quantities.

CHAPTER II.

DISCUSSION AND RESULTS OF OBSERVATIONS OF THE SUN.

Treatment of the Right Ascensions.

9. The meridian observations of the Sun have been treated on a system different in some points from that adopted in the case of the planets. It was possible to simplify the treatment by supposing that the small latitude of the Sun was always a definitely known quantity, so that when the observations were corrected for it the apparent motion of the Sun could be supposed to take place along the great circle of the ecliptic. This allowed the correction of the elements to depend on but two quantities—the obliquity of the ecliptic and the Sun's true longitude. Assuming the obliquity to be known, the longitude of the Sun could always be determined from an observation of its Right Ascension. An observed Right Ascension being compared with a tabular one, the residual gives rise to an equation of condition between the correction of the longitude, λ, of the obliquity, ε, and of the Right Ascension of the Sun, α:

$$d\alpha = \cos \varepsilon \sec^2 \delta d\lambda - \tfrac{1}{2} \tan \varepsilon \sin 2\alpha d\varepsilon.$$

This equation may be used to express the error of the longitude in terms of the error of the obliquity and of the Right Ascension as follows:

$$\delta\lambda = \sec \varepsilon \cos^2 \delta\delta\alpha + \tfrac{1}{2} \tan \varepsilon \sin 2\lambda d\varepsilon$$
$$= \sec \varepsilon \cos^2 \delta\delta\alpha + 0.21 \sin 2\lambda d\varepsilon$$

The elements mainly to be determined from the observations in Right Ascension being the eccentricity and perihelion of the Earth's orbit, each of the coefficients of which go through a period in a year, the effect of the small term $- 0.21\, \delta\varepsilon \sin 2\lambda$ whose coefficient does not amount to $0''.10$ after 1800, and has a period of half a year, will be practically without influence

15

on the result. The system was therefore adopted of deriving the residual in longitude directly from the residual in Right Ascension by the formula

$$\delta\lambda = F\delta\alpha$$

where

$$F = \cos^2 \delta \sec \varepsilon.$$

The residual $\delta\lambda$ in true longitude is then to be expressed in terms of the residual $\delta l''$ in mean longitude and of corrections to the eccentricity and to the longitude of the perigee relative to the Stars. In this expression the coefficient of the residual in mean longitude was always taken as unity, the value of the correction being so small in the case of LEVERRIER'S tables that no appreciable error would result from this supposition. Thus each residual in Right Ascension would give rise to an equation of condition of the form—

$$\delta l'' + Pe''\delta\pi'' + E\delta e'' = \delta\lambda = F\delta\alpha$$

We are here to regard $\delta l''$ and $\delta\pi''$ as corrections to the Right Ascensions relative to the clock stars, and not to the Sun's longitude or perigee simply. I shall therefore use the symbol c instead of $\delta l''$ to express the relative correction hereafter.

Treatment of the Declinations.

10. The declination of the Sun in the case supposed is a function only of the longitude and obliquity. The equation for expressing the observed correction in Declination in terms of the corrections to these two quantities is

$$\Delta\delta = \sin \alpha\delta\varepsilon + \cos \alpha \sin \varepsilon\delta\lambda$$

Thus each observation of the Sun's Declination gives rise to an equation of condition of this form.

It is however to be supposed that the observations in Declination made at each observatory will be affected by a constant error. If the observations are truly reduced to the standard system of star places, this error will be that of the standard system. As a matter of fact, however, observations made in the daytime, especially on the Sun and at noon, are made under circumstances so different from night observations on

stars that we can not assume the error of the reduced declination to be necessarily the same as that of the star system.
We must, therefore, in each case, regard the constant error in
declination as something peculiar to the observatory and the
instrument, which may or may not be worthy of subsequent
discussion. Thus each residual in declination gives rise to
an equation of condition,

$$\varDelta\delta_0 + \cos \alpha \sin \varepsilon\delta\lambda + \sin \alpha\delta\varepsilon = \varDelta\delta$$

$\varDelta\delta$ being the excess of observed over tabular declination,
and $\varDelta\delta_0$ the common error of all the measured declinations of
any one series.

Formation of the equations from Right Ascensions.

11. The method of treating the observed Right Ascensions
of the Sun was suggested by the fact that they are peculiarly
liable to systematic and personal errors; the former likely to
change with the seasons, and to be different for different instruments; and the latter to continue through the work of one
observer. It is now well understood that the observed Right
Ascensions of the mean of the Sun's two limbs relative to the
fixed stars are affected by personal errors, no means of eliminating which have yet been tried. In a series of observations
made by a single observer, under uniform conditions, this error
would systematically affect only the relative mean of the Right
Ascensions of the Sun and Stars, leaving the eccentricity and
perigee derived from the observations substantially correct.

On taking up the work it was also supposed that, owing to
the different effect of the Sun's rays upon the instrument at
different seasons, and the different circumstances under which
observations were made, the Right Ascensions of the Sun
would be affected by errors varying in a regular way through
the year, but not wholly expressible as a term of single annual
period. It was therefore deemed best to consider the observations possibly affected by an error of double period, having the
form

$$x' \cos 2g + y' \sin 2g$$

The introduction of the coefficients x' and y' added two more terms to the equations of condition, which terms, however, did not express any astronomical fact, but only the possible errors of the observations.

An additional and very important element to be determined from the observed Right Ascensions was the mass of Venus. The question now arose whether, by a uniform series of observations, extending through some definite period, the corrections to the eccentricity and perigee and the coefficients x' and y' could be completely separated from the coefficients of the correction to the mass of Venus. Examination showed that from such a series of observations, extending through eight years, the mass of Venus could be determined irrespective of all systematic errors repeating themselves with the season, provided that the observations were equally distributed throughout the year, or even that an equal number were made at the same time through successive years. As neither of these conditions are practically fulfilled it was judged best to assume in the beginning that the systematic errors of an unknown kind repeated themselves at each season during an eight-year period, and that they could be expressed in the form

$$c + x \cos g + y \sin g + x' \cos 2g + y' \sin 2g$$

x and y would appear as errors of eccentricity and perigee which could not be eliminated.

The quantities actually introduced as the unknown ones of the equations of condition were as follows:

μ', the factor of correction of the mass of Venus;

x, one-fifth the correction to the eccentricity;

y, one-fifth the correction $e'' \delta \pi''$;

x', y', one-tenth the coefficients expressing the supposed error of double period arising from all causes whatever;

c, the constant correction to the Right Ascension of the Sun relative to the Stars.

The coefficient of c was supposed unity throughout. The reduction of the residual in Right Ascension to that in Longitude and the other factors were taken from a table like the following, of which the argument was the day of the year.

Separate tables were constructed for 1802 and 1850, but they were so nearly identical that no distinction need be made between them. Furthermore, the error introduced by supposing the mean anomaly to have the same value on the same day of every year is entirely unimportant.

Table of coefficients for expressing errors of the Sun's Right Ascension in terms of errors of the elements of the Earth's orbit.

	$\dfrac{da}{dl}$	$\dfrac{dl}{du}$	Coefficients of—			
			$x = 0.2\delta e$	$y = 0.2e\delta\pi$	x'	y'
Jan. 1	1.09	0.91	+ 0.1	−10.0	+ 0.1	+10.0
11	1.07	0.93	1.8	9.8	3.5	9.4
21	1.04	0.96	3.4	9.4	6.5	7.6
31	1.01	0.98	5.0	8.7	8.7	5.0
Feb. 10	0.98	1.01	− 6.4	7.7	9.8	+ 1.8
20	0.96	1.04	+ 7.6	− 6.5	+ 9.9	− 1.6
Mar. 2	0.94	1.06	8.6	5.1	8.7	4.9
12	0.92	1.08	9.4	3.5	6.6	7.5
22	0.92	1.08	9.8	1.9	3.7	9.3
Apr. 1	0.93	1.07	10.0	− 0.1	+ 0.3	10.0
11	0.94	1.05	+ 9.9	+ 1.6	− 3.1	− 9.5
21	0.96	1.03	9.5	3.2	6.1	7.9
May 1	0.99	1.01	8.8	4.8	8.4	5.4
11	1.02	0.98	7.8	6.2	9.7	− 2.2
21	1.05	0.95	6.6	7.5	9.9	1.2
31	1.07	0.93	+ 5.3	+ 8.5	− 8.9	− 4.5
June 10	1.09	0.91	3.7	9.3	6.9	7.2
20	1.10	0.91	2.1	9.8	4.1	9.1
30	1.09	0.91	+ 0.4	10.0	− 0.7	10.0
July 10	1.08	0.93	− 1.3	9.9	+ 2.7	9.6
20	1.05	0.95	− 3.0	+ 9.5	+ 5.8	− 8.2
30	1.03	0.97	4.6	8.9	8.2	5.7
Aug. 9	1.00	1.00	6.1	8.0	9.6	+ 2.7
19	0.97	1.03	7.3	6.8	10.0	− 0.8
29	0.95	1.05	8.4	5.4	9.1	4.1
Sept. 8	0.93	1.07	− 9.2	+ 3.9	+ 7.2	− 6.9
18	0.92	1.08	9.7	2.3	4.5	8.9
28	0.92	1.08	10.0	+ 0.6	+ 1.2	9.9
Oct. 8	0.93	1.07	9.9	− 1.1	− 2.2	9.7
18	0.95	1.05	9.6	2.8	5.4	8.4
28	0.97	1.02	− 9.0	− 4.4	− 7.9	− 6.1
Nov. 7	1.00	0.99	8.1	5.9	9.5	− 3.1
17	1.03	0.96	7.0	7.2	10.0	+ 0.3
27	1.06	0.94	5.6	8.3	9.3	3.7
Dec. 7	1.08	0.92	4.1	9.1	7.5	6.6
17	1.09	0.91	− 2.5	− 9.7	− 4.9	+ 8.7
27	1.09	0.91	− 0.8	−10.0	− 1.6	+ 9.9

Finally, throughout the work the equations of condition were expressed only in entire numbers, the decimals being neglected. To lessen the number of equations of condition, the residuals were divided into groups generally covering from ten to fifteen days, the length of the group being determined by the condition that the perturbations of Venus must not change much during the period.

While the formation and solution of the equations of condition on this system were going on, it was found that the introduction of the assumed coefficients x' and y' was a refinement productive of little or no good result. In fact, the observations of the Sun proved to be much freer from annual sources of error than I had supposed, as will be seen by the tables of their results soon to be given. This is shown by the general consistency of the corrections to the eccentricity and perigee given by the work at the same or different observatories during different periods.

In marked contrast to this is the discordance among values of the correction c to the relative Right Ascensions of the Sun and Stars. This quantity it is that is affected by personal error and possibly by the effect of the Sun on the instrument. Under a perfect system of discussion it would be advisable to determine it separately for each observer. This however was practically impossible.

Solution of the equations.

12. For the purposes of forming and solving the normal equations, the equations of condition were divided into groups of generally from four to eight years, the exact lengths of which will be seen from the following exhibit of results. The equations for each period were solved on the supposition that the corrections were constant during the period. Thus every separate result is independent of every other, except so far as they may depend on the same instrument or the same observer at different times.

The first column shows the years through which the observations extend.

The second one shows to the nearest year the value of T— that is, the fraction of the century after 1850.

The third column shows the value of μ', or that factor which, being multiplied by the adopted mass of Venus, is to be applied as a correction to that mass, to obtain the value given by the observations.

All systematic errors arising from the instrument and the observer are so completely eliminated from the separate determinations of μ' that they may be regarded as absolutely independent of each other, that is—as not affected by any common systematic error.

We have next the relative weight assigned to each value of μ', which is determined in the usual way from the solution, and is, therefore, on a different scale for different observatories.

Next is given the value of c, or the apparent correction to the Right Ascension of the Sun, relative to the assumed Right Ascensions of the Stars, as given by observations during the several periods and expressed in seconds of arc, followed by the weights assigned to the separate results.

The next two columns, the corrections to the solar eccentricity and to the longitude of the perigee, require no further explanation.

Respecting the weights ultimately assigned to these quantities, and to c, it is to be remarked that they are the result of judgment more than of computation. It is only possible to enumerate in a general way with some examples the considerations on which they are based.

In assigning the weight of c the number of observers engaged is an important factor in determining it. Other factors are the steadiness of the atmosphere and the adaptation of the instrument to this particular work. General consistency is an important factor in the assignment. In this respect the Cambridge observations are quite remarkable ; if their excellence corresponds to their consistency they must be the best ones made.

It will be seen that PIAZZI's results are thrown out entirely. The wide range of his values of c led to the inquiry whether more consistent results would be obtained by taking shorter periods, but it was found that the values of c varied from time to time in such an irregular way that his instrument must have been affected by some extraordinary cause of error, unless some mistake has been made in interpreting or treating the observations.

The Oxford values of c are unusually discordant. The presumption that this discordance arises mainly from the special personal equation in observations of the Sun, described on page 17, derives additional weight from the greater relative consistency of the values of $\delta e''$ and $e'' \delta \pi''$. I have therefore allowed the values of these quantities to receive a fair weight.

The value of c for Paris, 1866–'70, has received a much reduced weight, solely on account of its excessive value. It seems that the work of one observer who made many observations during this period was affected by an unusual systematic error.

Results of observations of the Sun's Right Ascension.

GREENWICH.

Years.	T	μ'	w	c	w	$\delta e''$	$e'' \delta \pi''$	w
1750–'62	−.94	−.027	20	+0.33	1.5	+0.04	−0.42	2
1765–'71	−.82	−.041	10	+0.37	0.5	−0.08	−0.64	1
1772–'78	−.75	−.022	10	+0.74	0.5	−0.16	−0.49	1
1779–'85	−.68	−.035	5	+2.89	0.2	−0.18	−0.73	0.5
1786–'92	−.61	−.037	8	+1.51	0.2	−0.12	−0.88	0
1793–'97	−.55	−.114	5	+1.87	0.2	−0.22	−1.27	0
1798–'02	−.50	+.060	5	+1.02	0.2	−0.42	−1.15	0
1803–'06	−.45	−.002	5	+0.27	0.2	−0.03	−1.03	0
1807–'10	−.41	−.068	5	−0.34	0.2	−0.32	−1.12	0
1811–'14	−.37	−.095	3	−3.33	0.2	+0.17	−1 08	0
1815–'18	−.33	−.052	6	−1.99	0.5	−0.12	−0.34	0
1819–'22	−.29	+.010	6	−0.51	1	+0.22	−0.19	1
1823–'26	−.25	−.054	6	−1.08	1	+0.05	−0.17	1
1827–'30	−.21	−.045	6	−0.42	1	−0.09	−0.75	1
1831–'34	−.17	+.016	7	+0.76	1	+0.04	−0.27	1
1835–'38	−.13	+.020	8	+1.16	1	+0.26	+0.06	2
1839–'42	−.09	+.061	8	+0.84	1	+0.32	+0.10	2
1843–'46	−.05	−.008	8	+0.15	2	+0.25	+0.22	2
1847–'50	−.01	−.045	8	−0.10	2	+0.28	+0.02	3
1851–'54	+.03	+.024	8	+0.40	3	+0.22	+0.02	3
1855–'58	+.07	−.032	9	+0.36	3	+0.15	+0.02	3
1859–'62	+.11	−.043	9	−0.02	3	+0.25	+0.22	4
1863–'66	+.15	−.016	8	+0.31	3	+0.23	−0.05	4
1867–'70	+.19	+.031	8	+0.35	3	+0.33	−0.10	4
1871–'74	+.23	+.021	8	+0.12	3	+0.24	+9.05	4
1875–'78	+.27	−.008	8	−0.12	3	+0.26	+0.06	4
1879–'82	+.31	+.017	8	−0.05	3	+0.21	+0.14	4
1883–'88	+.36	+.001	13	−0.20	3	+0.18	+0.07	4
1889–'92	+.41	−.025	8	−0.44	2	+0.24	+0.11	3

Results of observations of the Sun's Right Ascension—Continued.

PARIS.

Years.	T	μ'	w	c	w	δe''	e''δπ''	w
1801-'07	−.46	−.025	14	−1.78	0.5	+0.08	−0.23	1
1808-'15	−.38	+.015	17	−0.65	0.5	−0.01	+0.12	1
1816-'22	−.31	−.050	14	+0.18	0.5	−0.13	+0.32	1
1823-'29	−.24	−.050	10	+0.01	0.5	−0.31	−0.02	1
1837-'44	−.09	−.034	19	+0.33	1	−0.04	+0.10	1.5
1845-'52	+.01	+.009	15	+0.10	1	+0.04	+0.10	1.5
1853-'59	+.06	+.014	15	+0.66	1	−0.04	+0.32	2
1860-'65	+.13	+.003	10	+0.38	1	+0.07	+0.26	2
1866-'70	+.18	.000	7	+2.29	0.3	+0.13	+0.40	2
1871-'79	+.25	+.048	11	−0.26	1	−0.06	+0.22	2
1880-'89	+.35	+.002	14	+0.44	1	+0.24	+0.03	2

PALERMO.

Years.	T	μ'	w	c	w	δe''	e''δπ''	w
1791-'96	−.56	−.079	0	−0.07	0	−0.06	−0.85	0
1797-'01	−.51	−.116	0	−2.33	0	−0.29	−0.28	0
1802-'05	−.46	−.001	0	−3.11	0	−0.05	−0.76	0
1806-'12	−.41	+.243	0	+5.92	0	−1.17	+1.55	0

CAMBRIDGE.

Years.	T	μ'	w	c	w	δe''	e''δπ''	w
1828-'34	−.21	+.007	16	−0.13	2	+0.08	+0.12	4
1835-'40	−.12	−.033	14	−0.18	2	+0.06	−0.06	4
1842-'47	−.05	−.026	9	−0.21	2	+0.08	−0.12	4
1850-'58	+.04	−.024	20	−0.11	2	+0.17	−0.04	4

WASHINGTON.

Years.	T	μ'	w	c	w	δe''	e''δπ''	w
1846-'52	−.01	−.038	5	−0.85	2	+0.20	0.00	3
1861-'65	+.13	−.038	8	−0.53	4	+0.01	0.01	5
1866-'73	+.20	−.004	13	−0.22	4	+0.18	−0.03	6
1874-'81	+.28	−.033	12	−0.45	4	+0.07	−0.16	5
1882-'91	+.37	−.002	17	−0.79	4	+0.07	−0.07	5

KÖNIGSBERG.

Years.	T	μ'	w	c	w	δe''	e''δπ''	w
1816-'23	−.30	+.002	13	+0.30	1	+0.07	−0.28	3
1824-'30	−.23	−.006	12	+0.02	1	−0.16	+0.11	3
1831-'38	−.15	−.021	15	+0.23	1	−0.12	+0.03	3
1839-'45	−.08	−.021	12	+0.77	1	+0.08	+0.20	3

Results of observations of the Sun's Right Ascension—Continued.

OXFORD.

Years.	T	μ'	w	c	w	$\delta e''$	$e''\delta\pi''$	w
1840–'49	—.05	—.043	12	+2.49″	0.3	+0.24″	—0.17″	2
1860–'68	+.14	+.042	13	+1.96	0.3	+0.08	—0.13	2
1869–'76	+.23	+.054	15	+0.92	0.3	+0.20	—0.04	2
1880–'87	+.34	—.014	9	—0.31	0.3	+0.27	+0.64	2

PULKOWA.

1842–'50	—.04	+.047	11	+1.20″	1	—0.12″	+0.20″	3
1861–'70	+.16	+.002	10	—0.40	1	+0.05	+0.28	3

DORPAT.

1823–'30	—.23	+.021	9	+0.36″	1	—0.12″	—0.22″	2
1831–'38	—.15	+.008	6	+0.45	1	+0.02	+0.03	2

CAPE OF GOOD HOPE.

1884–'90	+.37	—.026	12	—0.36″	3	+0.02″	+0.01″	4

STRASSBURG.

1883–'88	+.36	—.014	12	—1.65″	2	+0.23″	+0.09″	3

The mass of Venus.

13. The mean results for the mass of Venus given by the work at the several observatories are shown as follows:

The probable error, where given at all, is that derived from the discordance of the separate individual results at the particular observatory. In some cases there are only one or two results; here no probable error could be assigned.

w' is the sum of the weights of the result at each separate observatory, as given by the equations of condition. Were all the observations of equal accuracy, these would be the weights to be assigned to the separate results. Such not being the case, we choose for the actual weights certain numbers, founded partly on a compromise between the mean errors following each result or upon the values of w', partly on a judgment of the accuracy of the observations.

Values of μ' for the mass of Venus.

	μ'	W'	w
Greenwich	$-.015\pm.006$	226	11
Paris	$-.007\pm.009$	146	5
Königsberg	$-.012\pm.010$	52	3
Cambridge	$-.018\pm.009$	59	6
Dorpat	$+.016$	15	1
Pulkowa	$+.025$	21	1
Oxford	$+.014\pm.023$	49	1
Washington	$-.018\pm.009$	55	4
Cape	$-.026$	12	1
Strassburg	$-.014$	12	1

Using the weights in the last column, we have for the mean result

$$\mu = -.0118 \pm .0034.$$

The mean error $\pm.0034$ is that given by the discordance of the separate results of the preceding table.

Corrections of relative Right Ascensions.

14. The true values of the remaining quantities c, $\delta e''$, and $e''\delta\pi''$ are to be regarded as increasing uniformly with the time and therefore of the form

$$x + T\,y.$$

Here T is the time, and in the treatment of these particular equations it is counted from 1850 in units of one century, so that x is the value of the correction at this mean epoch.

The quantity designated by c is the same which, elsewhere in this discussion, is represented by $\delta l + \alpha$, so that

$$c = \delta l'' + \alpha$$

I shall, however, for convenience, continue to use the designation c, or $x+T\,y$.

As the observations at Greenwich and Paris extend over longer periods than at any other observatories, I shall first solve them separately. The totality of the Greenwich observations give for c the following normal equations and solution:

$$43.4\,x + 1.65\,y = +\,4''.23$$
$$1.65 + 4.24 \quad = -\,1''.25$$

$$x = +\,0''.11$$
$$y = -\,0''.34$$

Those at Paris give the equations and solution

$$8.3\,x + 0.04\,y = +\,1''.22$$
$$0.04 + 0.48 \quad = +\,0''.77$$

$$x = +\,0''.14$$
$$y = +\,1''.59$$

If we combine all the other results into a single set of normal equations, we have

$$40.2\,x + 4.26\,y = -\,10''.84$$
$$4.26 + 2.20 \quad = -\,3''.98$$

$$x = -\,0''.10$$
$$y = -\,1''.62$$

It will be seen that the results for y, the secular motion, are markedly discordant. Indeed, if we refer to the exhibit of results, p. 23, we shall see that the values of c are much more discordant than those of the other two quantities. To obtain a definite value, founded on all the observations of the Sun's Right Ascension, I do not see that any better result can be obtained than that found from a general solution of the combined normal equations. The equations and their solution are as follows:

$$91.9\,x + 5.95\,y = -\,5''.39$$
$$5.95 + 6.92 \quad = -\,4''.46$$

$$x = -\,0''.02$$
$$y = -\,0''.63$$

or

$$\delta l'' + \alpha = -\,0''.02 - 0''.63\,T$$

Corrections to the solar eccentricity and perigee.

15. I have already mentioned the remarkable consistency of the corrections to these elements given by the results at different observatories and at different epochs. The eccentricity is more consistent than the perigee. One cause for this, the consideration of which will throw some light on the relative merits of the observations, is that the error of Right Ascension depending on the Declination of the object observed effects the eccentricity less than the perigee. It is well known, from a comparison of the results, that the systematic differences in the Right Ascensions of different star catalogues vary somewhat with the Declination. Now, since the Sun's Declination goes through an annual period, it follows that this error will produce a systematic effect on both the eccentricity and the perigee. But the effect will be much larger in the case of the latter element than in the case of the former, because of the nearness of the perigee to the winter solstice, the difference being only some $10°$ or $12°$. Consequently the extreme coefficients in the correction to the eccentricity have nearly the same values, with opposite signs, for the same Declinations in different seasons of the year. But it is different with the perigee. The coefficient of this quantity is negative from October until March, when the Sun is in south Declination, attaining its maximum value about January 1; while it is positive during the remaining months when the Sun's Declination is north, attaining its maximum value about July 1. A systematic difference in the errors of Right Ascension will therefore produce its full effect on the longitude of the perigee, while its effect on the eccentricity will be but slight.

In this connection, the very large negative values of the correction to the perigee during the period when the old Greenwich transit instrument was in use are quite remarkable. The progressive change in the value of c is also remarkable in this connection. It is to be remarked that the new transit was mounted in 1816, but account was not taken of this fact in grouping the equations. Hence it is only from the year 1819 that the results of the table are derived wholly from observations with the new instrument. The anomaly alluded to is then seen to disappear. The fact that the abnormally large corrections in c are positive before 1800 and negative after it, while $e'' \delta \pi''$ is abnormally negative through the doubtful period 1765–1815, complicates the theory of these errors. I have not been able to consider them in detail, but have simply rejected the results for $\delta e''$ and $e'' \delta \pi''$ from 1786 to 1818, having given them a gradually diminishing weight from BRAD-LEY's observations to the first epoch.

As in the case of c, I have made a solution for Greenwich alone, Paris alone, the other observatories combined, and all combined. The results are shown as follows:

1. From Greenwich observations:

$$\begin{array}{cc} & \delta e'' \qquad\quad e'' \delta \pi'' \\ 54.5x + 2.73y = & + 11''.14; \; - 0''.88 \\ 2.73 \; + 5.72 \;\;\; = & + \;\; 1''.82; + 2''.69 \\ x = & + \;\; 0''.19; \; - 0''.04 \\ y = & + \;\; 0''.22; + 0''.49 \end{array}$$

2. From Paris observations:

$$\begin{array}{cc} & \delta e'' \qquad\quad e'' \delta \pi'' \\ 17.0x + 0.39y = & + 0''.30; + 2''.95 \\ 0.39 \; + 0.99 \;\;\; = & + 0''.29; + 0''.33 \\ x = & + 0''.01; + 0''.17 \\ y = & + 0''.29; + 0''.27 \end{array}$$

3. The equations and results from all the other modern observations are—

$$\begin{array}{cc} & \delta e'' \qquad\quad e'' \delta \pi'' \\ 77.0x + 4.99y = & + 5''.58; + 0''.35 \\ 4.99 \; + 3.68 \;\;\; = & + 1''.09; + 0''.40 \\ x = & + 0''.06; \;\; 0''.00 \\ y = & + 0''.22; + 0''.05 \end{array}$$

4. Finally, if we combine all the equations, we have—

$$\delta e'' \qquad e'' \delta \pi''$$

$$148.5x + \;8.11y = +\;17''.02; \; +\;2''.42$$
$$8.1\;\; +10.39\;\; = +\;\;3''.20; \; +3''.42$$
$$x = +\;\;0''.10; \qquad 0''.00$$
$$y = +\;\;0''.23; \; +\;0''.33$$

In the case of the eccentricity the general accordance is quite satisfactory, and for the perigee it is much better than in the case c, the relative Right Ascension.

Results of observed declinations of the Sun.

16. The Sun's absolute longitude can be found only from observations of his declination, because this longitude is referred to the equinox, which is defined only by the Sun's crossing of the equator.

The corrections to the eccentricity and perigee, as just found, are so slight that they may be neglected in determining the correction of the absolute longitude from that of the declination. Thus, as already stated, the unknown quantities of the equations given by the declinations are the corrections of the mean longitude l'', and of the obliquity ε, and a constant $\Delta\delta$, peculiar to each observatory, of which we take no further account. The equation of condition given by each observation or group of observations is

$$\Delta\delta + A \sin \varepsilon \delta l'' + B \delta \varepsilon = d\delta$$

where $d\delta$ is the excess of the observed over the tabular declination, and

$$A = \operatorname{cosec} \varepsilon \, \frac{d\delta}{d\lambda} = \cos \alpha$$

$$B = \qquad \frac{d\varepsilon}{d\delta} = \sin \alpha$$

The equations are grouped and solved for periods, as in the case of the Right Ascensions, with the results shown in the following table:

Results of observations of the Sun's Declination.

GREENWICH.

Years.	T	$\delta\iota''$	w	$\delta\varepsilon$	w	$\varDelta\delta$	$\delta'\varepsilon$	w
		''		''		''	''	
1753–'57	−.95	+0.78	1	−0.34	1	−2.43	−0.34	1
1758–'62	−.90	+1.50	1	−1.81	1	−1.94	−1.81	1
1765–'70	−.82	−0.23	1	−0.95	0.5	+0.20	−0.95	0.5
1771–'78	−.75	+0.48	1	−0.93	0.5	+1.25	−0.93	0.5
1779–'85	−.68	+1.23	1	−1.09	0.5	−0.99	−1.09	0.5
1786–'91	−.61	+0.48	1	−0.50	0.3	+0.15	−0.50	0.3
1792–'97	−.55	+1.12	1	−0.70	0.2	−0.35	−0.70	0.2
1798–'03	−.49	+0.41	1	−1.02	0.1	−0.10	−1.02	0.1
1804–'10	−.43	+0.18	1	−1.41	0.1	−0.84	−1.41	0.1
1812–'16	−.36	−0.15	3	−0.53	3	+0.48	−0.53	3
1817–'22	−.30	−0.41	3	+0.03	3	+0.40	+0.03	3
1823–'28	−.24	+0.43	3	−0.10	3	+0.08	−0.10	3
1829–'34	−.18	−0.08	3	+0.21	3	+0.25	+0.21	3
1835–'40	−.12	−0.12	3	−0.20	3	+0.37	−0.13	3
1841–'46	−.6	+0.21	3	+0.13	3	+0.47	+0.12	4
1847–'52	0	+0.25	4	0.00	4	−0.24	−0.15	4
1853–'58	+.6	+0.55	5	+0.18	5	−0.26	−0.05	5
1859–'64	+.12	+0.03	5	+0.28	5	−0.46	+0.12	5
1865–'70	+.18	−0.23	5	−0.15	5	+0.05	−0.36	5
1871–'76	+.24	−0.15	5	+0.26	5	+0.16	−0.16	5
1877–'82	+.30	−0.90	5	+0.22	5	+0.34	+0.08	5
1883–'88	+.36	−0.27	5	+0.33	5	−0.14	+0.02	5
1889–'92	+.41	−0.05	3	+0.19	3	+0.13	−0.07	3

PARIS.

Years.	T	$\delta\iota''$	w	$\delta\varepsilon$	w	$\varDelta\delta$	$\delta'\varepsilon$	w
1800–'03	−.48	+0.01	1	−1.93	1	−0.45	--------	------
1804–'07	−.44	+0.70	1	+0.82	1	−2.02	--------	------
1808–'10	−.41	+2.66	1	+1.60	1	−0.95	--------	------
1811–'15	−.37	−0.92	1	−1.20	1	−1.18	--------	------
1816–'21	−.31	+0.58	1	+1.68	1	−1.42	--------	------
1822–'28	−.25	+1.09	3	+0.39	3	−0.01	--------	------
1837–'42	−.10	+0.79	3	−0.15	3	+0.40	--------	------
1843–'48	−.4	+0.43	3	−0.03	3	+0.19	--------	------
1849–'54	+.2	+1.19	2	−0.01	2	+1.34	--------	------
1855–'60	+.8	+0.35	3	−0.02	3	+1.22	--------	------
1861–'66	+.14	+1.35	3	0.00	3	+0.12	--------	------
1867–'72	+.20	+0.31	2	−0.67	2	+0.10	--------	------
1873–'77	+.25	−0.59	2	+0.04	2	+1.01	--------	------
1878–'83	+.31	−0.09	2	−0.32	2	+0.58	--------	------
1884–'89	+.37	−0.80	2	+0.32	2	+0.78	--------	------

Results of observations of the Sun's Declination—Continued.

PALERMO.

Years.	T	δl″	W	δε	W	Δδ	δ'ε	W
		$''$		$''$		$''$	$''$	
1791-'03	—.53	—1.46	0	—0.95	------	+0.78	—0.95	0.4
1804-'13	—.41	+1.70	0	—0.52	------	+0.42	—0.52	0.4

CAMBRIDGE.

1833-'38	—.14	—0.21	2	—0.33	------	+0.59	—0.54	1
1839-'44	—.08	+0.31	2	—0.20	------	+0.29	—0.41	1
1847-'53	00	+0.21	2	+0.31	------	—0.32	+0.10	1
1854-'58	+.06	—0.15	2	+0.34	------	—0.42	+0.13	1

WASHINGTON.

1846-'49	—.02	—0.28	4	—0.73	------	—0.47	—0.81	2
1861-'66	+.14	—0.11	4	—0.43	------	—0.45	—0.25	2
1867-'72	+.20	+0.74	4	—0.39	------	+0.28	—0.51	2
1873-'78	+.26	—0.58	4	—0.32	------	+0.10	—0.45	2
1879-'84	+.32	—0.31	4	—0.60	------	—0.35	—0.72	2
1885-'91	+.38	—0.02	4	—0.05	------	—0.20	—0.18	2

KÖNIGSBERG.

1815	—.35	------	------	------	------	------	—1.07	0.5
1820-'23	—.28	—0.14	2	—0.22	------	—0.59	—0.47	1
1824-'27	—.24	+0.65	2	+0.49	------	—0.60	+0.24	1
1828-'31	—.20	+1.08	2	+0.09	------	—0.64	—0.16	1
1832-'34	—.17	—0.72	2	—0.15	------	—1.32	—0.40	1
1837-'44	—.09	—0.66	2	—0.62	------	—2.24	—0.87	1

OXFORD.

1840-'45	—.07	+0.79	2	+0.42	------	+0.67	+0.22	0.2
1846-'51	—.01	+0.35	2	+0.40	------	+0.89	+0.20	0.2
1861-'66	+.14	+0.36	2	—0.81	------	+0.10	—1.01	0.2
1867-'72	+.20	—0.16	2	—0.24	------	+0.29	—0.44	0 2
1873-'76	+.25	—0.38	2	—0.33	------	+0.29	—0 53	0.2
1880-'83	+.32	—0.43	2	+0.12	------	—0.17	—0 08	0.2
1884-'87	+.36	—0.24	2	+0.23	------	—0.19	+0.03	0.2

Results of observations of the Sun's Declination—Continued.

PULKOWA.

Years.	T	$\delta l''$	W	$\delta \varepsilon$	W	$\Delta \delta$	$\delta' \varepsilon$	W
		$''$		$''$		$''$	$''$	
1842–'45	−.06	+0. 82	2	−0. 35	------	−0. 01	−0. 35	1
1846–'49	−.02	−0. 10	2	−0. 48	------	+0. 07	−0. 48	1
1861–'65	+.13	−0. 53	2	−0. 48	------	−0. 30	−0. 48	1
1866–'70	+.18	+0. 27	2	−0. 31	------	−0. 38	−0. 31	1

DORPAT.

Years.	T	$\delta l''$	W	$\delta \varepsilon$	W	$\Delta \delta$	$\delta' \varepsilon$	W
1823–'28	−.24	+0. 99	2	−1. 26	------	+0. 59	−1. 41	1
1829–'32	−.19	+0. 99	2	−0. 76	------	+1. 34	−0. 91	1
1833–'38	−.14	+1. 00	2	−0. 63	------	+1. 34	−0. 78	1

CAPE OF GOOD HOPE.

Years.	T	$\delta l''$	W	$\delta \varepsilon$	W	$\Delta \delta$	$\delta' \varepsilon$	W
1884–'87	+.36	−0. 51	4	+0. 05	------	+0. 11	−0. 07	2
1888–'90	+.39	−0. 84	4	+0. 09	------	+0. 19	−0. 21	2

STRASBURG.

Years.	T	$\delta l''$	W	$\delta \varepsilon$	W	$\Delta \delta$	$\delta' \varepsilon$	W
1884–'88	+.36	−0. 57	4	−0. 05	------	−0. 77	+0. 12	2

LEIDEN.

Years.	T	$\delta l''$	W	$\delta \varepsilon$	W	$\Delta \delta$	$\delta' \varepsilon$	W
1864–'69	+.17	+0. 14	4	−0. 01	------	+0. 27	−0. 24	2
1870–'76	+.23	−0. 23	4	−0. 06	------	−0. 04	−0. 29	2

Correction to the Sun's absolute longitude

17. So far as mere instrumental measurement is concerned, the correction $\delta \varepsilon$ should be determined with greater precision than $\delta l''$ in the ratio 5:2, because the errors in declination have to be divided by the factor sin $\varepsilon = 0.40$, in order to form $\delta l''$. Allowing for this large increase in the source of error, the values of $\delta l''$ are more accordant than those of $\delta \varepsilon$. This is what we should expect. The values of the former quantity depend mainly upon the comparison of observations made

near the opposite equinoxes, when the sun has the same declination, and when the season is not greatly different. Indeed,
if the season changed exactly with the sun's declination, all
effects of annual change of temperature would be completely
eliminated from $\delta l''$, as would also in any case any constant
error which is a function simply of the Sun's Declination. It
is therefore to be expected that the actual probable error of
this quantity will conform more nearly to that determined from
the residuals than in the case of the other.

For these reasons the value of $\delta l''$ does not give rise to
much discussion. The general result from all the observatories is, for $\delta l''$, when developed in the form $x + y$ T.

$$x = + 0''.05$$
$$y = - 0''.97.$$

Obliquity of the ecliptic.

18. The determination of the obliquity rests upon an essentially different basis from that of the absolute longitude, in
that it depends upon actual differences of measured Declinations, which differences are still further complicated by the
fact that they are necessarily made at opposite seasons. A
more detailed discussion of them is therefore necessary, and
some modification may have to be made in the separate results
as adopted. The following special circumstances affecting the
observations are to be taken into consideration:

The BRADLEY Greenwich results for 1753–'62, are derived
from a manuscript communicated by Dr. AUWERS, containing
the results of his very careful reduction of BRADLEY's observed Declinations of the Sun, which were compared with
HANSEN's tables. The corrections were reduced to those of
LEVERRIER's tables by being computed at intervals sufficiently short to permit of the reduction being interpolated with
all necessary precision. No reduction was applied either on
account of the constant error of the Declinations determined
by Dr. AUWERS himself, nor for reduction to the BOSS system
of standard Declinations. Hence arises the large value of $\Delta\delta$
given by these Declinations. Consequently the value of $\delta\varepsilon$ is

that given immediately by the instrument, on the system of
reduction adopted by Dr. AUWERS, in which I have supposed
that the Pulkowa refractions were used.

From 1765 to 1816 the Greenwich observations were made
with the imperfect quadrant, the Declinations of which are
subjected to an error which is not constant. The necessary corrections are derived by SAFFORD in Vol. II of the
Astronomical Papers. The corrections are those necessary to
reduce to Boss's system, and they vary with the Declination.
Hence the arc on which the obliquity depends is not that
measured with the instrument itself, but that so corrected as
to reproduce as nearly as may be the standard Declinations.

From 1812 onward the two mural circles were used. Up to
1830 no correction except the constant one derived by SAF-
FORD was applied to the Declinations as measured with these
instruments. Hence the arc of obliquity is that measured
with the instrument itself without being corrected by the
standard stars.

After 1830 the Declinations were corrected by the tables for
Greenwich given in Boss's paper. These corrections vary
somewhat with the Declination, and they are different also
for different periods. Hence we have here a period during
which the instrumental differences of Declination were corrected to reduce them to the standard star-system.

If the standard system were subject to no further error than
a constant one, common to all Declinations within the zodiac,
which common correction would be subject to a uniform change
with the time, this system would doubtless be the best one to
adopt in order to obtain the secular variation in the obliquity
of the ecliptic. But, as a matter of fact, the standard Declinations are simply the mean results of Declinations measured
with different instruments. It is, therefore, a question whether
we shall get any better results by applying reductions to a
standard system than we should get by simply taking the
mean of the instrumental results, because the system is itself
only a mean of such results. It is true that the standard system depends on more instruments than the obliquity, though
not on better ones; but it is also to be considered that the
reductions in the case of the Sun may be different from those

in the case of the stars, owing to the very different conditions in which the observations are made.

Another troublesome point arises from the refraction used in the reductions. The effect of refraction is always to make the measured obliquity less than the actual one; the correction to the obliquity on account of refraction is therefore a positive quantity, which is a minimum for an observatory at the equator and increase equally towards each pole. Some values of the obliquity were derived from BESSEL's refractions of the *Tabulæ Regiomontanæ*, and others from the Pulkowa tables. Since the secular variation of the obliquity is more important than the absolute value of the quantity, it is essential that the standard to which all determinations of the obliquity are reduced should be as nearly as possible the same, and therefore that the same refraction should be used. But in reductions to standard star places we meet with the additional complication that the differences in the constant of refraction might be wholly or partially eliminated by the reductions to a standard system. It would therefore be a difficult question how far we should modify the values of $\delta\varepsilon$ on account of the use of different tables of refraction.

To avoid all these difficulties I have judged it best to make the obliquity depend mainly upon absolute measures, the reductions being made with the Pulkowa refractions.

Effect of refraction on the obliquity.

19. The determination of the average or most probable effect on the obliquity produced by using the Pulkowa refractions, instead of those of the *Tabulæ Regiomontanæ*, is easily determined. We divide the ecliptic into a number of equal arcs throughout the year, and by equations of condition express differences of refraction in terms of differences of Declination, and hence differences of obliquity. We thus find that at certain latitudes where observations were made, and where BESSEL's refractions were used in the reduction, the following corrections are necessary to reduce the obliquity to the ones given by the Pulkowa refractions:

$$\text{Pulkowa;} \qquad \varphi = 59°.8; \quad \Delta\varepsilon = -0''.325$$
$$\text{Greenwich;} \qquad \varphi = 51°.5; \quad \Delta\varepsilon = -0''.20$$
$$\text{Washington;} \quad \varphi = 38°.9; \quad \Delta\varepsilon = -0''.125$$

Hence I conclude that for

$$\text{Dorpat;} \quad \varDelta\varepsilon = -0''.29$$
$$\text{Königsberg;} \quad \varDelta\varepsilon = -0''.26$$
$$\text{Cambridge;} \quad \varDelta\varepsilon = -0''.21$$
$$\text{Cape Town;} \quad \varDelta\varepsilon = -0''.12$$

The corrections to the obliquity thus derived, depending mainly on direct instrumental measurement, and reduced to the Pulkowa refractions, are designated as $\delta'\varepsilon$. The results for this quantity are given in the last column of the several tables.

In the case of BRADLEY'S Greenwich results, I have taken as $\delta'\varepsilon$ Dr. AUWERS's results unchanged, assuming in the absence of any specific statement that he has used the Pultowa refraction tables.

In the case of MASKYLENE'S observations, I have, by exception, used them as reduced to the standard star-system, because we have no other results at these times, and the error of his instrument is so strongly shown that it would not do to use the results unchanged. It will be seen, however, that small weights are assigned, and that the weights diminish towards the end of the series.

In the case of the Greenwich observations from 1812 to about 1834, no change has to be made, as the results are generally or always purely instrumental, and Pulkowa refractions are used in SAFFORD's work.

From 1835 onward I have depended mainly on certain corrected Greenwich reductions. First, for $\delta'\varepsilon$, I have used the results given by Mr. CHRISTIE in his very valuable paper on the Greenwich Declinations, in M. R. A. S., Vol. XLV, where the Declinations from 1836 to 1879 are reduced on a uniform system. Later, I have adopted the corrected results given in Appendix III to the Greenwich observations for 1887. In each case the result has been reduced to the Pulkowa refractions.

The Paris results rest on a different basis from the others, in that the zero point of the instrument depends wholly upon LEVERRIER's Declinations of the stars, and I fear it was not always accurately determined. Observations near the winter solstice are mostly referred to one set of stars; those near the

summer to another set, the error of which may be systematically different. Certain it is that the results during the early
years were very discordant. The weights as given in the table
are those assigned *a priori*, without sufficient reference to the
discordance of the older results. I have felt constrained to
evade a decision as to their treatment by entirely omitting
their results in the final discussion.

In the case of some other observatories it was difficult to
determine exactly what refractions had been used in each
special case and what reductions should be made. I have, however, determined the corrections in the best way I was able.

A precise determination of the secular change in the obliquity is of more importance for our present object than a
precise determination of its amount. Hence a series of observations extending through a long period of time, and made on
a uniform system, has an advantage over a number of isolated
values, in that any constant error with which it may be
affected will be eliminated from the secular variation. Possible constant differences between the determinations of the
various observatories at different epochs will vitiate the secular variation, but the probable amount of this error may be
diminished by using a number of separate determinations,
such as are presented in the preceding table. In the Greenwich transit circle we have a very uniform series, extending
over a period of forty years, but giving results systematically
different from other determinations. This series gives for the
correction to the obliquity:

Transit Circle, 1847–'91:

$$\delta'\varepsilon = -0''.11 \pm 0''.06 + (0''.21 \pm 0''.46)\,T \quad . \quad . \quad . \quad (a)$$

Here, in view of the uniformity of method and reduction,
we may regard the mean error of the centennial variation from
the discordance alone as a fair approximation to the probable
mean error. It will be seen that I have here included four
years (1847–'50) of the Mural Circle results.

Continuing the Greenwich series backward, the question
arises whether we can regard the results of the mural circle
from 1812 to 1850 as comparable with those of the transit circle.

There is certainly nothing in the table to indicate any system-, atic difference. From the combination of the two we have—

M. C. and T. C., 1812–'50:

$$\delta'\varepsilon = -0''.08 \pm 0''.05 + (+0''.14 \pm 0''.23)\, T\, (1850) \quad .\quad . \quad (b)$$

Here the mean error is naturally smaller than in the case of the transit circle alone, but is now more subject to possible systematic difference between the two instruments.

If we now go back to BRADLEY, we meet with the very diffi- cult question, whether we should regard his results as best comparable with the modern Greenwich observations, or with modern observations in general. If we assume that the differ- ence between the Greenwich and other modern results is due to any cause which has remained unchanged since BRADLEY, we should reach one conclusion; otherwise, we should reach the other. The result of combining all Greenwich observa- tions, with the weights as assigned, is—

$$\delta'\varepsilon = -0''.11 + 0''.50\, T \quad . \quad . \quad . \quad . \quad (c)$$

In this combination I have used the weak results of MASKE- LYNE, with the small weights assigned, although they depend wholly upon the standard declinations of stars. In view of the discordance between BRADLEY'S two results, this seems the only admissible course.

Next in the length of time which they include come the Paris observations, of which the results, with the weights assigned, are—

$$\delta\varepsilon = +0''.01 - 0''.36\, T$$

I give this result in order that nothing may be omitted. Undue weight has probably been assigned to the earlier determinations; in any case the method of deriving it from the original observations is so objectionable that no further use is made of it. A satisfactory discussion of the observa- tions would require a complete redetermination of the zero points of the instrument from fundamental stars.

If we omit the Greenwich, Paris, and Palermo results, and combine all the others into a single set of equations of condition, we have the equations and results:

$$36.9\,x + 0.26\,y = -14''.37$$
$$0.26 + 1.88 = + 1''.01$$

$$x = -0''.39$$
$$y = +0''.59$$

Here x is the value of $\delta'\varepsilon$ for 1860, and y its centennial variation. Transferring the epoch to 1850, as usual, the result is—

$$\delta'\varepsilon = -0''.45 + 0''.59\,T \quad \cdots \quad (d)$$

No reliable mean error can be computed, owing to systematic errors. In view of these, one mode of treatment would be to form equations of condition in which a possible systematic error at each observatory would appear as one of the unknown quantities. By this process we should get the same result for the secular variation as if we made an independent determination from the work of each observatory. At most of the observatories the period through which the observations are made, with one instrument and on an unchanged plan, is too short to render such a course advisable.

As a last combination, we shall combine the earlier Greenwich results, up to 1810, with Palermo and with all the modern results except Paris, first dividing the weights of the Greenwich results by 2. We then have the equations—

$$39.8\,x - 1.82\,y = -17''.12$$
$$-1.8 + 3.47 = +2''.99$$

$$x = -0''.40$$
$$y = +0''.65 \quad \cdots \quad (e)$$

Concluded results for the obliquity.

20. The data on which these various results for the obliquity rest show the following noteworthy features:

(1) That the correction given by the modern Greenwich instruments, mural and transit circles, is markedly greater

than that given by other modern observations. This may be most plausibly attributed to the atmospheric conditions within the observing room.

(2) The minuteness of the change of the correction given by these instruments during nearly eighty years. To this circumstance is due the smallness of the centennial variation, $0''.50$, found from the totality of the Greenwich observations. A comparison of BRADLEY with the mean of the T. C. results only would have given a change of $0''.97$ in 117 years, or a centennial change of about $0''.80$.

The long period, uniformity of plan, and systematic deviation of the modern Greenwich observations lead me to consider them as forming a series distinct from all others. We have therefore the following two completely independent determinations of the centennial variation:

(1) Modern Greenwich results: $y = + 0''.14 \pm 0''.23$
(2) All other results $+ 0''.65$

To the latter no reliable mean error can be assigned. To judge its reliability we may compare it with the results (a), (c), and (d)—

Greenwich T. C., alone, $+ 0''.21 \pm 0''.46$
Greenwich observations in general, $+ 0''.50$
Miscellaneous modern observations, $+ 0''.59$

We may, it would seem, fairly give double weight to the result (2), thus obtaining, as the definite result from observations of the Sun alone:

Correction to LEVERRIER's centennial variation of the obliquity of the ecliptic ($- 47''.594$)

$$+ 0''.48 \pm 0''.30$$

the mean error being an estimate from the general discordance of the data.

For the constant part of the correction I take—

$$\delta\varepsilon\,(1850) = - 0''.30$$

Summary and comparison of results.

21. From what precedes we have the following as the values of the unknown quantities, and of their secular variations, as given by observations of the Sun alone.

	Value for 1850.	Cent. var.
$\delta e'' =$	$+ 0''.10 \pm 0''.03$	$+ 0''.23 \pm 0''.10$
$e''(\delta \pi'' + \alpha) =$	$0''.00 \pm 0''.07$	$+ 0''.33 \pm 0''.12$
$\delta l'' + \alpha =$	$- 0''.02$	$- 0''.63$
$\delta l'' =$	$+ 0''.05 \pm 0''.12$	$- 0''.97 \pm 0''.23$
$\delta \varepsilon =$	$- 0''.30 \pm 0''.15$	$+ 0''.48 \pm 0''.30$
$\alpha =$	$- 0''.07$	$+ 0''.34$

No estimate of the probable errors of these quantities would be useful which did not take account of the systematic differences between the results of different observatories. We have therefore formed the mean outstanding residual corrections given by the several observatories, as shown in the tables which follow. Originally the scale of weights used for the Greenwich observations did not correspond to that for the other observatories; they were, therefore, divided by 2. As used below, however, the change has been made in the case of $\delta l''$ by multiplying all the weights of the other observatories by 2, and, in the case of $\delta \varepsilon$, by dividing the Greenwich weights by 2.

The correction to the obliquity depends solely on $\delta' \varepsilon$; but the comparison has also been made with the values of $\delta \varepsilon$, which, it will be remarked, differ from the others in that account is taken of the supposed variation of the systematic correction with the declination. It is noteworthy that the results are somewhat more accordant when this correction is omitted and purely instrumental errors are used for the obliquity.

The mean errors given in the preceding summary of results are derived from the discordances in question, and may be regarded as substantially real.

No use was made of the Paris results for $\delta l''$ and $\delta \varepsilon$ for the reason that they depend on declinations referred to star places which may be affected by differences in different Right Ascensions. They are, however, retained in the table to show the amounts of outstanding discordance.

Outstanding mean residual corrections to quantities depending on the Sun's Right Ascension.

	$\delta e''$	$e''\delta\pi''$	Σw
Greenwich	$+ 0''.09$	$- 0''.03$	54.5
Paris	$- 0''.09$	$+ 0''.17$	17
Cambridge	$+ 0''.02$	$0''.00$	16
Washington	$- 0''.05$	$- 0''.12$	24
Königsberg	$- 0''.08$	$+ 0''.08$	12
Oxford	$+ 0''.06$	$+ 0''.02$	8
Pulkowa	$- 0''.15$	$+ 0''.22$	6
Dorpat	$- 0''.10$	$- 0''.03$	4
Cape	$- 0''.16$	$- 0''.11$	4
Strassburg	$+ 0''.05$	$- 0''.03$	3
Mean errors for weight unity $\varepsilon_1 =$	$\pm 0''.34$	$\pm 0''.39$	
Mean error of x	$\pm 0''.03$	$\pm 0''.03$	
Mean error of y	$\pm 0''.10$	$\pm 0''.12$	

Outstanding mean residual corrections to quantities depending on the Sun's Declination.

	$\delta l''$	w	$\delta \varepsilon$	w	$\delta'\varepsilon$
Greenwich	$- 0''.06$	64	$+ 0''.31$	29.6	$+ 0''.17$
Paris	$+ 0''.45$	0	$+ 0''.31$	0	
Palermo	$- 0''.39$	0	$- 0''.20$	0.8	$- 0''.20$
Cambridge	$- 0''.05$	8	$+ 0''.35$	4	$+ 0''.14$
Washington	$+ 0''.07$	24	$- 0''.22$	12	$- 0''.29$ ·
Königsberg	$- 0''.20$	10	$+ 0''.31$	5.5	$0''.00$
Oxford	$+ 0''.14$	14	$+ 0''.19$	1.4	$- 0''.01$
Pulkowa	$+ 0''.12$	8	$- 0''.13$	4	$- 0''.13$
Dorpat	$+ 0''.75$	6	$- 0''.49$	3	$- 0''.64$
Cape	$- 0''.35$	8	$+ 0''.10$	4	$- 0''.02$
Leiden	$+ 0''.10$	8	$+ 0''.17$	2	$- 0''.06$.
Strassburg	$- 0''.26$	4	$+ 0''.08$	4	$+ 0''.25$
ε for weight unity	$\pm 0''.81$		$\pm 0''.74$		$\pm 0''.60$

CHAPTER III.

RESULTS OF OBSERVATIONS OF MERCURY, VENUS, AND MARS.

Elements adopted for correction.

22. We first give an outline of the method of expressing the observed corrections to the Right Ascensions and Declinations of each of the planets as linear functions of the corrections to the tabular elements. This linear function forms the first member of the equation of condition in its original form, and the observed correction forms its second member.

Let us put—

    R, r, the radii vectores of the Earth and planet;

      L, the Sun's true longitude;

      J, the inclination of the orbit of the planet to a plane passing through the Sun's center parallel to the plane of the Earth's equator;

      N, the Right Ascension of the ascending node of the orbit on this plane;

      U, the argument of heliocentric declination of the planet or its angular heliocentric distance from the node on the equator;

      α, δ, the geocentric Right Ascension and Declination of the planet.

      ε, the obliquity of the ecliptic;

We shall then have—

$$\alpha = f(r.\ \mathrm{R.\ L.\ J.\ N.\ U.}, \varepsilon.) \quad . \quad . \quad . \quad . \quad . \quad . \quad . \quad . \quad (a)$$

For the correction to the tabular Right Ascension arising from symbolic corrections to these seven quantities, we have the equation—

$$\delta\alpha = \frac{d\alpha}{d\mathrm{J}}\delta\mathrm{J} + \frac{d\alpha}{d\mathrm{N}}\delta\mathrm{N} + \frac{d\alpha}{du}\delta\mathrm{U} + \frac{d\alpha}{dr}\delta r + \frac{d\alpha}{d\varepsilon}\delta\varepsilon$$
$$+ \frac{d\alpha}{d\mathrm{L}}\delta\mathrm{L} + \frac{d\alpha}{d\mathrm{R}}\delta\mathrm{R} \quad (1)$$

43

with a similar equation for the declination, formed from this by writing δ for α.

The relations by which these two equations are derived, as well as the expressions for the differential coefficients they contain, are given very fully in A. P., Vol. II, Part I, to which reference may be made. The corrections δN and δU are not, however, the most convenient ones to choose. It will be found in the paper alluded to that they have been transformed by measuring the longitude in orbit of the planet and that of the perihelion from an arbitrary point in the orbit. As to this very convenient device in celestial mechanics, it is to be remarked that the "departure point" always disappears from the final equations which determine the position of the planet. We may, in fact, make abstraction of it by considering that its introduction is equivalent to the following simple linear transformations.

We put

   w, the distance from the node to the perihelion;
   f, the true anomaly;
   g, the mean anomaly.
   π, the longitude of the perihelion;
   l, the mean longitude of the planet;
   v, its true longitude;

these longitudes being counted from the departure point.

Then we have the relations—

$$\delta U = \delta w + \delta f = \delta v - \cos J \delta N$$
$$\delta w = \delta \pi \ - \cos J \delta N \qquad\qquad (2)$$
$$\delta l = \delta \pi \ + \delta g$$

Hence,

$$\delta \pi = \delta U + \cos J \delta N - \delta f$$
$$\delta f = \frac{df}{dg} \delta g + \frac{df}{de} \delta e \qquad\qquad (3)$$

The elements finally adopted for correction by the equations of condition were—

$$l. \ \pi. \ e. \ J. \ N.$$

The value of a, the mean distance, is known with such precision that its correction need not enter into the equations of condition. The latter are formed by substituting in (1)

$$\delta U = \left(1 - \frac{df}{dg}\right) \delta \pi + \frac{df}{de} \delta e + \frac{df}{dg} \delta l - \cos J \delta N.$$
$$\delta r = \frac{dr}{de} \delta e + \frac{dr}{dg} \delta l - \frac{dr}{dg} \delta \pi \tag{4}$$

The coefficients of each equation of condition from the Right Ascension thus become—

$$\text{Coefficient of } \delta J \quad \cdots \quad \frac{d\alpha}{dJ}$$

$$\text{``} \quad \text{``} \quad \delta N \quad \cdots \quad \frac{d\alpha}{dN} - \cos J \frac{d\alpha}{du}$$

$$\text{``} \quad \text{``} \quad \delta\varepsilon \quad \cdots \quad \frac{d\alpha}{d\varepsilon}$$

$$\text{``} \quad \text{``} \quad \delta l \quad \cdots \quad \frac{d\alpha}{dU} \frac{df}{dg} + \frac{d\alpha}{dr} \frac{dr}{dg} \tag{5}$$

$$\text{``} \quad \text{``} \quad \delta\pi \quad \cdots \quad \frac{d\alpha}{dU}\left(1 - \frac{df}{d\pi}\right) - \frac{d\alpha}{dr} \frac{dr}{dg}$$

$$\text{``} \quad \text{``} \quad \delta e \quad \cdots \quad \frac{d\alpha}{dU} \frac{df}{de} + \frac{d\alpha}{dr} \frac{dr}{de}$$

In the second members of the equations α is regarded as a function of the seven quantities (a), as is also δ, for which a similar equation is to be formed.

The corrections of the solar eccentricity, perihelion, and mean longitude were also introduced by putting in (1)

$$\delta L = \delta l'' + \frac{dL}{de''} \delta e'' + \frac{dL}{d\pi''} \delta \pi''$$
$$\delta R = \qquad \frac{dR}{de''} \delta e'' + \frac{dR}{d\pi''} \delta \pi'' \tag{6}$$

Introduction of the masses of Venus and Mercury.

23. The correction to the mass of Venus was introduced by taking the tabular perturbation produced by Venus on the geocentric place of the planet at the mean date of each equation as the coefficient of the unknown quantity to be determined. In computing these perturbations regard was

had to the action of Venus on the Earth as well as on the planet. On this system the unknown quantity finally found would be the factor by which the adopted mass of the planet must be multiplied in order to give the correction of that mass.

It has already been remarked that the mass of a planet can not be determined free from systematic error by observations made upon the planet itself. Hence, the mass of Venus can be determined only from observations of Mercury and Mars, and that of Mercury only from observations of Venus and Mars. But the mass of Mercury is so minute that it would be useless to attempt to determine it from observations either of the Sun or Mars. It was therefore determined solely from the periodic perturbations of Venus.

It has happened that the mass of Venus could not be determined in a reliable way from observations of Mars, owing to a defect in the theory of the latter planet, which I shall mention hereafter, and have not yet had time to correct. Practically, therefore, the mass of Venus is determined only from observations of the Sun and of Mercury, and that of Mercury from observations of Venus.   .

Correction of equinox and equator.

24. Could all the observations be directly referred to a visible equinox and equator, the corrections above enumerated would have been the only ones which it was necessary to include in the equations of condition. But, as a matter of fact, the observations were all referred to an assumed system of Right Ascensions and Declinations of standard stars—my own system in Right Ascension and Boss's in Declination. We must therefore introduce two additional unknowns into the equations, which I have represented in the following way:

α, the common error of the adopted Right Ascensions.

δ, the common error of Boss's Declinations.

The first quantity will appear only in the equations derived from observed Right Ascensions and the second only in the equations derived from Declinations, the coefficient being unity in each case.

That the value of δ found in this way should be regarded as a correction to the Declinations of the equatorial stars will appear by the following considerations. The mean heliocentric orbit of a planet as projected on the celestial sphere is undoubtedly a great circle. On the other hand, in view of the systematic discordance always found to exist in measures of absolute Declinations near the equator, and of the fact that these absolute Declinations depend upon assumed constants and laws of refraction, which are necessarily affected with greater or less uncertainty, and are otherwise subject to systematic errors, instrumental or personal, of an obscure character, but strongly shown by a comparison of the Declinations derived from the work of different observatories, it can not be assumed that these Declinations are free from systematic error. Now, in one circle of Declination, say the equator, we may expect that the error will be nearly constant around the sphere, since the causes of error will generally be nearly constant for any one Declination. This conclusion is confirmed by a comparison of the best star catalogues. Moreover, between the zodiacal limits, the error in each particular case is not likely to differ very greatly from the error at the equator. Even if the difference should be considerable the various values of the error of the different Declinations must have a certain mean value, so that in the case of each particular star, or each region of the heavens, we may conceive the actual error to be divided into two parts—one the mean value in question, and the other the deviation from this mean. The latter is probably smaller than the former, and in any case can not very well be determined from observations of the planets. But the condition that the planet moves on a great circle of the sphere admits of the mean value being determined with great precision. It should, therefore, be included in the equations of condition.

The value of α, the common error of all the Right Ascensions, can obviously not be determined from the equations in Right Ascension alone, because the only result that such observations can give us would be the values of the Right Ascensions referred to some assumed equinox. The coefficient of α would therefore completely disappear from the equations of condition in Right Ascension. But since the same unknown quantities are introduced into the equations of condition in Right Ascension and in Declination, the requirement that the two sets of equations shall give common values of these quantities does away with this indetermination and enables determinate values to be found. In fact, this method does not differ in principle from that usually adopted in deriving the Right Ascensions of stars from observations of the Sun. The latter consists in deriving the Sun's absolute longitude from observations of its Declination and absolute Right Ascensions· of the stars by comparing them with the Sun. In the same way we may.consider that, in observations of the planet, the Sun's absolute longitude is derived from observations of Declinations of the planet, and then α comes out from the observations in Right Ascension.

I have deemed it absolutely necessary that all the equations of condition should be solved by the method of least squares. By this method alone can the results of the observations as regards separate values of the elements and constants be properly brought out. But the work of constructing and solving a system of nine thousand equations of condition, each involving twenty unknown quantities, would be extremely laborious, and might even require a century for its completion, if done in the usual way. It was therefore necessary to adopt every device by which the labor could be reduced to a minimum. One device was the dropping of all superfluous decimals in the coefficients of the equations. Since the errors thus produced would be purely accidental, it follows that if the sum of the products obtained by multiplying the value of each unknown quantity by the error of its coefficient in the equation of condition is but a small fraction of the necessary probable error of the absolute term, no serious harm will result from the errors of the coefficients.

Another device was the construction of tables for finding the coefficients. Such tables relating to Mercury and Venus are found in Vol. II, Part 1, of the *Astronomical Papers*. These tables are, however, only given for one mean anomaly in each case, and therefore require computations dependent on the value of the other anomaly. They were therefore extended to tables of double entry, so that the value of the derivatives of the geocentric Right Ascension or Declination at any epoch could be taken from the tables at sight. The arguments were the mean anomaly of the planet and the day of the year at which the planet last passed through its perihelion.

Introduction of the secular variations.

25. When the equations of condition are formed on the plan just set forth, the unknown quantities will be the corrections to the elements or to the mean longitude at the date of each equation. But every one of the unknown quantities which have been enumerated, the correction of the masses excepted, is subject to a secular variation. Hence, instead of the unknown quantities heretofore defined, we introduce two others, the one the value of this unknown at some assumed mean epoch, which, for reasons already set forth, must first be determined from the observations; the other the secular variation in a unit of time. The unknown quantities which have been enumerated make twelve for each equation of condition. Eleven of these are subject to a secular variation, so that if the secular variations were introduced into the original equations of condition they would each have twenty-three unknown quantities.

The following device was employed to reduce to a minimum the work of introducing and determining the secular variations of the various elements:

Firstly, the whole time covered by the observations was divided into periods, never exceeding ten years, except when the observations were very few in number, or entitled to but small weight. It was then assumed that no error would arise from supposing the value of the unknown quantity to be the same throughout the period as it was at the mid-epoch of the period. The maximum absolute error thus arising would be the secular variation during half the length of the period, and the mean error the secular variation during one-fourth of the period; but actually the effect of even this error would be almost entirely nullified by the combination of positive and negative coefficients throughout each period.

5690 N ALM——4

Let us now put

$$x, y, \quad . \quad . \quad .$$

the corrections to the elements at any epoch, τ.

Let

$$a\,x + b\,y + c\,z + \quad . \quad . \quad . = n$$

be an equation of condition between these quantities at this epoch. From a system of such equations, extending through a period numbered i, during which x, y, etc., may be considered as constant, we derive normal equations of the form—

$$[aa]_i\,x + [ab]_i\,y + \quad . \quad . \quad . = [an]_i$$
$$[ab]_i\,x + [bb]_i\,y + \quad . \quad . \quad . = [bn]_i$$
$$. \quad . \quad . \quad . \quad . \quad . \quad . \quad . \quad . \tag{1}$$

which I shall call partial normal equations, and which we might solve so as to obtain the values of x, y, etc. This solution is not, however, necessary. The values of the unknown quantities being really of the general form—

$$x = x_0 + x'\,t$$
$$y = y_0 + y'\,t \tag{2}$$
$$. \quad . \quad . \quad . \quad . \quad .$$

we may imagine these values substituted in the normal equations (1), the value τ_i of t for the mean epoch of the period being substituted for t.

Let us now suppose that we introduce the quantities $x_0, y_0, \ldots,$ $x', y', \ldots$ into the original equations of condition, using for t the value τ_i, which pertains to the mean epoch of the period. Our equation of condition will thus become—

$$ax_0 + by_0 + \quad . \quad . \quad + a\tau_i x' + b\tau_i y' + \quad . \quad . \quad = n \tag{3}$$

If from a system of conditional equations of this form we form the normal equations for all the unknown quantities, the results will be these:

Partial normal equation in x_0;

$$[aa]_i x_0 + [ab]_i y_0 + \quad . \quad . + \tau_i [aa]_i x' + \tau_i [ab]_i y' + \quad . \quad . = [an]_i \tag{4}$$

Partial normal equation in x';

$$\tau_i [aa]_i x_0 + \tau_i [ab]_i y_0 + \; . \; . \; + \tau_i^2 [aa]_i x' + \tau_i^2 [ab]_i y'$$
$$+ \; . \; . \; = \tau_i [an]_i \quad (5)$$

We conclude that the partial normal equations, when the full number of unknown quantities is included, may be derived from those of the form (1) by the following rules.

(1) Each partial normal equation in x_0, y_0, . . . is formed from that in x, y, etc., by adjoining to the first member of the equation the member itself multiplied by τ and then changing x, y, . . . to x_0, x_0; and, in the products by τ, changing x, y, . . . into x', y', . . .

(2) The partial normal equation in x', y', . . . is formed from the partial equation in x_0, y_0, . . . by multiplying all the terms throughout by the factor τ.

The final or complete normal equations in all the unknown quantities being formed by the addition of the partial normals, the formulæ for the coefficients are as follow:

For the final equation in x_0

$$
\begin{aligned}
[aa] &= [aa]_1 + [aa]_2 + \; . \; . \; . \; + [aa]_n \\
[ab] &= [ab]_1 + [ab]_2 + \; . \; . \; . \; + [ab]_n \\
& \;\; . \;\; . \qquad . \;\; . \qquad . \;\; . \qquad\qquad . \;\; . \\
[aa]' &= \tau_1 [aa]_1 + \tau_2 [aa]_2 + \; . \; . \; . \; + \tau_n [aa]_n \\
& \;\; . \;\; . \; . \qquad . \;\; . \; . \qquad . \;\; . \; . \qquad\qquad . \;\; . \; . \\
[an] &= [an]_1 + [an]_2 + \; . \; . \; . \; + [an]_n
\end{aligned}
\quad (6)
$$

For the final equation in x'

$$
\begin{aligned}
[aa]'' &= \tau_1^2 [aa]_1 + \tau_2^2 [aa]_2 + \; . \; . \; . \; + \tau_n^2 [aa]_n \\
[ab]'' &= \tau_1^2 [ab]_1 + \tau_2^2 [ab]_2 + \; . \; . \; . \; + \tau_n^2 [ab]_n \\
& \;\; . \;\; . \qquad . \;\; . \qquad . \;\; . \qquad\qquad . \;\; . \\
[an]'' &= \tau_1 [an]_1 + \tau_2 [an]_2 + \; . \; . \; . \; + \tau_n [an]_n
\end{aligned}
\quad (7)
$$

The final equations for all the unknown quantities will then be of the form

$$
\begin{aligned}
&[aa] \; x_0 + [ab] \; y_0 + \; . \; . \; + [aa]' \; x' + . \; . \; . = [an] \\
&\quad . \; . \; . \qquad . \; . \; . \qquad\qquad . \; . \; . \qquad\qquad . \; . \; . \\
&[aa]' x_0 + [ab]' y_0 + . \; . \; . + [aa]'' x' + . \; . \; . = [an]''
\end{aligned}
\quad (8)
$$

The epoch from which we count the time, τ, is arbitrary. An obvious advantage will be gained in counting it from the mid-epoch of all the observations. Then we shall have, by putting w_1, w_2, etc., for the sum of the weights for the different periods:

$$w_1 \tau_1 + w_2 \tau_2 + \quad . \quad . \quad . \quad + w_n \tau_n = 0 \qquad (9)$$

If the observations are then equally distributed around the orbits of the planet and of the Earth it may be expected that the coefficients

$$[aa]', \ [ab]' \quad . \quad . \quad . \quad . \qquad (10)$$

will all nearly or quite vanish. Practically we may expect that as observations are continued through successive revolutions the ratios of these to the other coefficients will approach zero as a limit. We may then divide the normal equations into two sets, one containing the quantities x_0, y_0, etc., and the other x', y', etc. The coefficients (10) being small, the two sets of normals will be nearly independent, and we may omit the terms (10) in the first approximation, and introduce them in one or two successive approximations so far as necessary.

The unit of time is also arbitrary. A certain advantage in symmetry will be gained by so choosing it that the mean value of τ^2 shall not differ greatly from unity. It was found that twenty-five years was a sufficiently near approximation to be adopted for all three planets.

Dates and weights for epochs and periods.

26. As want of space makes impracticable the present publication of the great mass of material worked up, the following particulars have been selected as those most likely to be useful in judging and criticising the work. We give three tables, showing the division of the dates of observation into periods, and the weights for each period. The first column of each table contains the number or designation of the period, as found in the manuscript books. The second contains the mean year of the period. The third column shows the time

of this mean period from the mid-epoch of the observations, which is taken as follows:

For Mercury, 1865.0

Venus,   1863.0

Mars,   1856.0

The next column contains the sum of the weights of the equations in each period, as used in forming the normal equations. These were not, however, the weights actually used in multiplying the coefficients of the equations of condition. Owing to the diversity in the quality of the observations at different times it was not found convenient to reduce the equations at once to a uniform system of weights, and so different units of weight were selected for the older observations and for the earlier observations. After the partial normal equations were formed they were multiplied by the factor F, necessary to reduce them to a standard in which the unit of weight should correspond to the mean error—

$$\varepsilon_1 = \pm\ 1''.0$$

The sums of the weights reduced by these factors are shown in the table.

In arranging the weights and selecting the factors it should be remarked that a liberal allowance was made at each step for probable constant errors, which results in the given weights being much smaller than they would have been by the theoretical treatment of the original observations. Notwithstanding this allowance the final result seems to show that it was still insufficient, and that the actual weights of the results are less than would follow even from the final ones as given.

The partial normal equations for each period after being multiplied by the factors F, are added to form the final normal equations as derived from meridian observations.

Weights, epochs, and periods of partial normal equations.

MERCURY.

Period.	Right Ascension.				Declination.			
	Mean year.	τ (units of 25 *y.*)	Wt.	F.	Mean year.	τ (units of 25 *y.*)	Wt.	F.
1	1766.60	−3.9360	3.4	1/8	1765.50	−3.9800	0.2	1/8
2	1784.22	−3.2312	18.8	1/8	1782.99	−3.2804	4.9	1/8
3	1799.81	−2.6076	26.1	1/8	--------	------------	------	----
3_1	--------	------------	------	----	1796.42	−2.7432	5.0	1/8
3_2	--------	------------	------	----	1802.37	−2.5052	39.9	3/10
4	1809.53	−2.2188	18.9	1/8	1809.18	−2.2328	52.8	3/10
5	--------	------------	------	----	1824.83	−1.6068	74.1	3/10
5_1	1818.79	−1.8484	0.9	1/8	--------	------------	------	----
5_2	1825.80	−1.5680	34.5	1/4	--------	------------	------	----
6	1835.56	−1.1776	75.0	1/4	--------	------------	------	----
6_1	--------	------------	------	----	1833.84	−1.2464	75.3	3/10
6_2	--------	------------	------	----	1838.26	−1.0696	141.5	1/2
7	1843.74	−0.8504	98.8	1/4	1843.97	−0.8412	281.5	1/2
8	1855.90	−0.3640	83.3	1/4	1855.92	−0.3632	201.5	1/2
9_1	1863.10	−0.0760	99.8	1/4	1862.79	−0.0884	189.5	1/2
9_2	1867.12	+0.0848	186.0	1/4	1867.18	+0.0872	294.5	1/2
10_1	1872.62	+0.3048	129.8	1/4	1872.64	+0.3056	214.0	1/2
10_2	1877.12	+0.4848	129.8	1/4	1877.05	+0.4820	204.5	1/2
11_1	1882.24	+0.6896	108.2	1/4	1882.17	+0.6868	171.5	1/2
11_2	1886.29	+0.8516	199.8	1/4	1886.29	+0.8516	338.0	1/2
11_3	1889.82	+0.9928	109.5	1/4	1889.70	+0.9880	176.0	1/2

VENUS.

Period.	Mean year.	τ (units of 25 *y.*)	Wt.	F.	Mean year.	τ (units of 25 *y.*)	Wt.	F.
1	1755.83	−4.2868	11.3	1/8	1759.69	−4.1324	7.0	1/4
2	1767.92	−3.8032	19.7	1/8	1770.18	−3.7128	10.0	1/4
3	1781.06	−3.2776	3.7	1/8	1793.25	−2.7900	13.5	1/4
4	1792.47	−2.8212	12.3	1/8	1806.73	−2.2508	65.5	1/2
5	1802.64	−2.4144	23.3	1/8	1815.59	−1.8964	67.5	1/2
6	1810.31	−2.1076	34.0	1/8	1823.75	−1.5700	197.0	1
7	1816.88	−1.8448	42.7	1/8	1836.02	−1.0792	762.0	1
8	1825.55	−1.4980	141.0	1/8	1844.08	−0.7568	650.0	1
9	1835.31	−1.1076	339.3	1/8	1854.24	−0.3504	333.0	1
10	1843.98	−0.7608	259.3	1/8	1861.43	−0.0628	749.0	1
11	1853.51	−0.3796	205.3	1/8	1868.06	+0.2024	815.0	1
12	1861.60	−0.0560	353.7	1/4	1875.32	+0.4928	692.0	1
13	1868.12	+0.2048	466.0	1/2	1883.15	+0.8060	819.0	1
14	1875.38	+0.4952	399.5	1/2	1888.56	+1.0224	801.0	1
15	1883.09	+0.8036	514.5	1/2	--------	------------	------	----
16	1888.67	+1.0268	520.5	1/2	--------	------------	------	----

Weights, epochs, and periods of partial normal equations.

MARS.

Period	Right Ascension.				Declination.			
	Mean year.	r (units of 25 y.)	Wt.	F.	Mean year.	r (units of 25 y.)	Wt.	F.
1	1757.43	−3.9428	25.3	⅓	1758.82	−3.8872	8.8	⅓
2	1770.55	−3.4180	11.0	⅓	1773.79	−3.2884	8.8	⅓
3	1787.82	−2.7272	10.0	⅓	1794.48	−2.4608	13.0	⅓
4	1799.77	−2.2492	20.7	⅓	1804.91	−2.0436	47.0	⅓
5	1811.32	−1.7872	14.7	⅓	1813.00	−1.7200	30.5	⅓
6	1829.17	−1.0732	60.0	⅓	1828.04	−1.1184	93.0	1
7	1837.39	−0.7444	121.0	⅓	1837.18	−0.7528	371.0	1
8	1845.39	−0.4244	76.3	⅓	1844.95	−0.4420	255.0	1
9	1853.36	−0.1056	90.0	⅓	1853.02	−0.1192	245.0	1
10	1861.07	+0.2028	114.0	⅓	1860.94	+0.1976	306.0	1
11	1869.20	+0.5280	124.0	⅓	1868.80	+0.5120	197.0	1
12	1877.71	+0.8684	132.0	½	1877.38	+0.8552	257.0	1
13	1883.27	+1.0908	91.0	⅓	1883.26	+1.0904	160.0	1
14	1888.85	+1.3140	115.5	⅓	1888.48	+1.2992	167.0	1

Unknown quantities of the equations.

27. For convenience in solving the equations of condition the coefficients of the equations were multiplied by such numerical factors as would reduce their general mean absolute value to numbers of approximately the same order of magnitude. Hence, the unknown quantities themselves are not the corrections to the elements, but these corrections divided by the adopted factors.

In the case of Mercury the absolute term was also multiplied by 10, so that effectively the factors in question were reduced to one-tenth part of their value. The unknown quantities of the equations are represented by the symbols of the elements to which they relate inclosed in brackets.

For convenience of reference the following table is given, showing the factors used in the case of each planet. In the case of Mercury the column (*a*) shows the factors by which the differential coefficients were actually multiplied; (*b*) the factor by which the unknown quantity, as finally found, must be multiplied to obtain the correction as expressed in the last column.. In the case of Venus and Mars these factors are the same.

Factors by which the unknown quantities are to be multiplied to obtain corrections of the elements.

Symbol of unknown.	Factor for—				Corr. of element.
	Mercury.		Venus.	Mars.	
	(a)	(b)			
[m]	1	0.1	7	0.3	$\delta m : m_0$
[l]	40	4	5	2	δl
[J]	30	3	6	2.5	δJ
[N]	30	3	7	2.5	$\sin J \delta N$
[e]	30	3	3	$10 \div 7$	δe
[π]	100	10	439	$100 \div 7$	$\delta \pi$
[π]	100	2.056	3	1.3323	$e \delta \pi$
[ε]	10	1	4	4	$\delta \varepsilon$
[e'']	6	0.6	2.5	2	$\delta e''$
[π'']	6	0.6	2	2	$e'' \delta \pi''$
[α]	10	1	1	5	α ·
[δ]	10	1	5	5	δ
[l'']	10	1	4	3	$\delta l''$

The secular variation of each unknown in 25 years is expressed sometimes by a suffixed 1, sometimes by an accent, thus:

$$[l]' = [l]_1 = \text{change of } [l] \text{ in 25 years.}$$

28. It may also be useful to give the values of the principal coefficients in each of the normal equations. They are found in the following table. Were the other coefficients all zero, these numbers would indicate the weights of the different unknown quantities as resulting from the solution. Several of them were greatly diminished by the process of solution.

Values of the principal diagonal coefficients in the normal equations.

Symbol of coefficient.	Mercury.			Venus.			Mars.
	From mer. observations.	From transits.	Sum.	From mer. observations.	From transits.	Sum.	From mer. observations.
mm	5488	0	5488	5868	2929	8797	17887
ll	10559	11308	21867	5981	3540	9521	20924
JJ	15222	1296	16518	13232	7444	20676	28783
NN	14176	2304	16480	17951	1636	19587	32478
ee	19015	5076	24091	5686	3350	9036	20119
$\pi\pi$	8621	8352	16973	5290	1732	7022	20564
$\varepsilon\varepsilon$	11001	196	11197	11429	3598	15027	31460
$e''e''$	9757	508	10265	9586	665	10251	15909
$\pi''\pi''$	9099	261	9360	5836	1895	7731	14911
$\gamma''\gamma''$	5242	0	5242	----	----	----	----
$l''l''$	13041	542	13583	11031	2349	13380	15427
aa	13230	0	13230	335	0	335	25138
$\delta\delta$	24657	0	24657	15196	0	15196	53975
ll'	7014	67155	74169	6005	8983	14988	26689
JJ'	12366	9383	21749	9837	13014	22851	23440
NN'	11035	16682	27717	14724	2874	17598	29494
ee'	15437	29647	45084	5743	8610	14353	24364
$\pi\pi'$	6745	49318	56063	4948	4483	9431	27131
$\varepsilon\varepsilon'$	8488	1418	9906	8458	6306	14764	25675
$e''e''\,'$	8409	2937	11346	9805	1682	11487	22947
$\pi''\pi''\,'$	8439	1513	9952	5242	4805	10047	17356
$\gamma''\gamma''\,'$	5432	0	5432	----	----	----	----
$l''l''\,'$	11629	3126	14755	10677	5667	16344	20655
aa'	11400	0	11400	297	0	297	33624
$\delta\delta'$	18716	0	18716	10772	0	10772	42405

NOTE.—The coefficients for Mercury and Venus in this table are given as they were used in the solution, after dropping the units from all the terms of the equations, except those from transits of Mercury.

Order of elimination.

29. In dealing with so extensive a system of unknown quantities it is impracticable to investigate the dependence of each upon all the others. It is therefore essential to arrange the unknowns in an order partly that of interdependence and partly that of the liability of each to subsequent change by discussion and adjustment. Hence, the mass of the planet, Mercury or Venus, should be first eliminated, as being that unknown which is least affected by changes in the final values of the other unknowns. The secular variations, as derived

from meridian observations, are nearly independent of the corrections to the other elements. The solar elements are to be subsequently determined by a combination of the results of the observations of the Sun and of the three planets.

Guided by these considerations, the order of elimination was, with some exceptions, as follows:

1. The mass of the disturbing planet.
2. The five elements of the observed planet.
3. The four elements of the Earth's orbit.
4. The corrections to the star-positions for the mid-epoch.
5. The secular variations of the eleven quantities (2), (3), and (4), taken in the same order.

Treatment of meridian observations of Mercury.

30. In the case of Mercury the factors of the coefficients of the equations were chosen large enough to admit of the decimals being dropped from the products without prejudice to the accuracy of the final result. This was done to facilitate the formation of the normal equations. For the same reason the factors were made so small that the absolute numerical values of the coefficients should generally not exceed 13. As this degree of precision is far short of that usually employed for correcting the elements of a planet, it may be well to set forth the considerations on which it is based.

Let any equation of condition as actually used be—

$$ax + by + cz + \ .\ .\ .\ = n \qquad (a)$$

Let the coefficients a, b, etc., be affected by the mean errors ε, ε', etc., so that the true equation should be—

$$(a + \varepsilon) x + (b + \varepsilon') y + \ .\ .\ .\ = n$$

This true equation may be written in the form—

$$ax + by + \ .\ .\ .\ = n - \varepsilon x - \varepsilon' y - \ .\ .\ . \qquad (b)$$

We may regard (b) as a rigorous equation, in which the error of the second member is increased by the quantity—

$$\pm\, \varepsilon x \pm \varepsilon' y \pm.$$

and the only effect upon the precision of the results will be
that arising from this increased probable error. Let us estimate its magnitude. From an examination of the tables used
in finding the coefficients I infer that the probable error of the
coefficient of π was ± 1, and that of all the other coefficients
± 0.6. The mean value of the unknown quantities was generally a small fraction of a second. We conclude, therefore,
that the probable or mean value of the error

$$\pm \, \epsilon x \pm \epsilon y \pm \quad . \quad . \quad .$$

would in any case be only a small fraction of a second. Moreover, these errors would be purely accidental and not systematic, since the intervals of time between the equations were
generally so long that the coefficients for different equations
came from different tables, so that no error from omitted decimals in any one equation would enter into the other equations.

Now, in view of the necessary systematic errors which affect
observations of the planets, there is no hope of approximating
to this degree of accuracy in the second members of the equations. Were the observations rigorously correct and the
values of the unknown quantities finally determined affected
by no error except that arising in this way, they would be
many times more accurate than we can hope to make them.
The errors might, in fact, be considered unimportant in the
present state of astronomy.

It has already been remarked that the scale of weights was
so taken that the unit of weight should correspond approximately to a supposed mean error $\pm 1''.0$ in the value of each
absolute term of an equation of condition, so far as the error
could be determined from the discordance of the original
observations. The corresponding probable error would be
$\pm 0''.65$. In the case of Mercury, however, modifications were
made which prevents this mean error from corresponding to
the unit of weight which would be found from the solutions in
the usual way. In the first place, the absolute members were
all multiplied by 10; in other words, the decimal point was
dropped from tenths of seconds, and no further account taken
of it. Secondly, in consequence of the probable error in the
coefficients of the normal equations arising from the imperfec-

tions of the decimals, the final values of these coefficients would be subject to probable errors ranging between 50 and 100 units. In consequence there would be no advantage in retaining the last figure in the normal equations, and it was dropped in all the subsequent solution and discussion of these equations.

In dropping the last figure from the absolute term of the normal equations we may consider that we are merely dropping the tenths of seconds and that the units are once more expressed in seconds. Thus, considering only the effect of this operation, the unit of weight would correspond to a mean error of ± 1.0 in units of the absolute term. But in dropping off the last figure from the coefficients we practically reduce the scale of weights, considered as multipliers of the equations, to one-tenth of their former value. On the other hand, in expressing the unknown quantities in terms of the corrections to the elements, we divide the multipliers by ten, so that effectively we multiplied the coefficients in the equations of condition, considering the unknown quantities to be defined as on page 56, by 10. Since these coefficients are of the second degree in the normal equations, it follows that the scale of weights has in effect been increased ten fold. Hence the unit of weight for the normal equations between the unknown quantities as finally solved will correspond to the mean error

$$\varepsilon_1 = 1.0 \times \sqrt{10} = \pm 3.1$$

As the mean error is at best a rather indefinite quantity in a case like the present, we may consider its value as 4 units and even then as by no means rigorously determined.

Up to the time of writing no attempt has been made to derive rigorously the weights of the unknown quantities from the solution, because in the cases of most of the unkowns such weights would be entirely illusory. The fact is that in solving so immense a mass of equations, we must expect systematic errors to vitiate many of the results. The observations of Mercury, especially of its Right Ascension, are not made on a uniform system; sometimes the limb is observed, sometimes the apparent center or the center of light.

An ideally perfect system of reduction would require us to reduce each separate observation with a semidiameter corresponding to the personal equation of the observer. This being entirely impracticable, we must regard the reduction of the observer's semidiameter to that used in the reductions as a probable error. In fact, however, it will be of a systematic character, varying at each point of the relative orbit of Mercury, and going through a cycle of changes impossible to determine in a synodic period of the planet. It is impracticable to give even a full discussion of these errors; we shall, however, meet with a proof of their magnitude.

Introduction of the equations derived from observed transits of Mercury.

31. The relations between the elements of Mercury and the Earth derived from this source are shown in my *Discussion of Transits of Mercury* (A. P., Vol. I, Part VI.) On page 447 are found expressions for those linear functions of the corrections to the elements which are determined by the November and May transits, respectively. With a slight change of notation to correspond with that of the present paper, these functions are as follows:

$$V = 1.487\,\delta l - 0.487\,\delta\pi - 1.137\,\delta e - 1.01\,\delta l'' + 1.19\,e''\delta\pi''$$
$$+ 1.58\,\delta e''$$
$$W = 0.716\,\delta l + 0.284\,\delta\pi + 0.896\,\delta e - 0.97\,\delta l'' - 1.11\,e''\delta\pi''$$
$$- 1.62\,\delta e''$$

The values of V and W being derived from a series of transits extending from 1677 to the present time, enable us to determine both these quantities at some epoch, and their secular variations. The values derived from the transits, together with their mean errors, are found on page 460 of the work in question. Omitting the doubtful factor k, introduced on account of a possible variability of the Earth's axial rotation, which was not proved by the transits, the values of V and W were found to be as follows:

$$V = -0''.90 \pm 0''.31 + (-2''.63 \pm 0''.59)\,(T - 1820)$$
$$W = +0''.84 \pm 0''.25 + (+1''.84 \pm 0''.60)\,(T - 1820) \quad (a)$$

The mean epoch for the transits is taken as 1820, to which the zero values correspond. The values for 1865.0, the mid epoch for the meridian observations, are, therefore, from the transits alone— ·

$$V = - 2''.08 \pm 0''.41$$
$$W = + 1''.67 \pm 0''.37$$

This, however, is only a first approximation to the quantities which should be introduced. Since the meridian observations help to determine the values of V' and W', we should not regard the reductions to 1865.0 as final, but retain the results in the form (a).

Another element which is determined from the observed transits of Mercury with greater precision than it can be from meridian observations is the longitude of the node of the orbit relatively to the Sun. In the paper quoted we have put—

$$N = (\delta\theta - \delta l'') \sin i$$

and found from all the transits up to 1881,

$$N = - 0''.16 \pm 0''.27 + (0''.28 \pm 0''.62)(T - 1820) \qquad (b)$$

The values of V, W, and N, found from the discussion in question, give rise to six conditional equations, which become completely independent when we take as observed values the secular motions and the absolute values at the mid-epoch of observation. This mid-epoch is not the same for the May and November transits. But I have assumed that no serious error would be introduced by taking 1820.0 as the epoch for all three of the quantities, V, W, and N.

If we substitute for $\sin i \, \delta\theta$ its value in terms of δJ, etc., namely,

$$\operatorname{Sin} i \, \delta\theta = - 0.6018 \, \delta J + 0.796 \sin J \delta N + 0.721 \, \delta\varepsilon \qquad (c)$$

and then for δJ, δN, $\delta\varepsilon$, their values in terms of the unknowns of the equations of condition, we shall have

$$N = - 1.805 \, [J] + 2.394 \, [N] + 0.721 \, [\varepsilon] - 0.122 \, [l''] \qquad (d)$$

Similar expressions will be found for the values of V and W by substituting for the corrections to the elements the unknown quantities of the conditional equations, as already given.

Taking 1820.0 as the mid epoch, we may regard the independent quantities given by the transits of Mercury to be the six following ones:

$$V_0 - 1.8\,V_1 \; ; \; W_0 - 1.8\,W_1 ; \; N_0 - 1.8\,N_1 \tag{e}$$
$$V_1 \quad\; ; \quad W_1 \quad\; ; \quad N_1 \; \cdot$$

Here V_0, W_0, and N_0 indicate values for 1865, the mid-epoch of the meridian observations; and V_1, W_1, and N_1 the variations in 25 years. The six conditional equations thus found from the transits may be written

$$V_0 - 1.8 \; V_1 = - 0''.90 \pm 0''.31$$
$$W_0 - 1.8\,W_1 = + 0''.84 \pm 0''.25$$
$$N_0 - 1.8 \; N_1 = - 0''.16 \pm 0''.27$$
$$V_1 \qquad\;\; = - 0''.66 \pm 0''.15 \tag{f}$$
$$W_1 \qquad\;\; = + 0''.46 \pm 0''.15$$
$$N_1 \qquad\;\; = + 0''.07 \pm 0''.15$$

Substituting for V_0, V_1, etc., their expressions as linear functions of the unknowns of the conditional equations, we find the following six equations, which are to be used as conditional equations additional to those given by the meridian observations:

$$5.95\,[l] - 4.87\,[\pi] - 3.41\,[e] - 1.01\,[l''] + 0.71\,[\pi''] + 0.95\,[e'']$$
$$-1.8\{5.95[l]_1 - 4.87\,[\pi]_1 - 3.41\,[e]_1 - 1.01\,[l'']_1 + 0.71\,[\pi'']_1$$
$$+ 0.95\,[e'']_1\} = - 0''.90$$
$$\text{Weight} = 250$$

$$2.86\,[l] + 2.84\,[\pi] + 2.69\,[e] - 0.97\,[l''] - 0.67\,[\pi''] - 0.97\,[e'']$$
$$-1.8\{2.86\,[l]_1 + 2.84\,[\pi]_1 + 2.69\,[e]_1 - 0.97\,[l'']_1 - 0.67\,[\pi'']_1$$
$$- 0.97\,[e'']_1\} = + 0''.84$$
$$\text{Weight} = 300$$

$$- 1.8\,[J] + 2.4\,[N] + 0.7\,[\varepsilon] - 0.12\,[l'']$$
$$-1.8\{ - 1.8\,[J]_1 + 2.4\,[N]_1 + 0.7\,[\varepsilon]_1 - 0.12\,[l'']_1\} = - 0''.16$$
$$\text{Weight} = 400$$

$$5.95\,[l]_1 - 4.87\,[\pi]_1 - 3.41\,[e]_1 - 1.01\,[l'']_1 + 0.71\,[\pi'']_1 + 0.95\,[e'']_1$$
$$= -0''.66$$

Weight = 700

$$2.86\,[l]_1 + 2.84\,[\pi]_1 + 2.69\,[e]_1 - 0.97\,[l'']_1 - 0.67\,[\pi'']_1 - 0.97\,[e'']_1$$
$$= +0''.46$$

Weight = 700

$$-1.8\,[J]_1 + 2.4\,[N]_1 + 0.7\,[\epsilon]_1 - 0.12\,[l'']_1 = +0''.07$$

Weight = 1,600

The weights assigned to these several equations have been determined by the following considerations:

We have already found that in the equations of condition from the meridian observations as finally reduced, the scale of weights has so come out as to show a practical mean error for weight unity of about $\pm\,4''$. Were this error purely accidental, the weights of the conditional equations derived from the transits would be determined in the same way, from the mean errors assigned to them. But, as a matter of fact, the existence of systematic errors in the meridian observations is shown, as will be subsequently explained, by the large value found for the fictitious quantity δr_2. Since observations of transits are made at the point of the relative orbits of Mercury and the Earth, near which meridian observations are rarely available, and are of a higher order of accuracy than meridian observations, it follows from the theory of probabilities that we should assign a larger relative weight to the observations of the transits. How much larger does not admit of being determined with numerical precision. Actually I have taken the weights as if the mean error corresponding to weight unity were between 5 and 6. In the case of the motion of the node a still larger weight has been assigned to the secular variation, from the belief that the accuracy of the determination from transits relative to meridian observations is in this case of a yet higher order of magnitude than in the case of the other elements. Whether this belief is justified or not must be left to the decision of the future astronomer.

The first three of the preceding six conditional equations may be treated in a way similar to that adopted for the meridian observations. They express what is supposed to be equivalent to observations of the three quantities V, W, and N in 1820, when $\tau = -1.8$. Hence, from the partial normals in the six principal unknowns, $[e], [\pi] \ldots [e'']$, the complete normals may be formed by multiplication by τ and τ^2 $(\tau = -1.8)$ in the way set forth in § 25.

Solutions of the equations for Mercury.

32. In the case of Mercury and Venus, it is desirable to know to what extent the results of the transits diverge from those of the meridian observations. Hence, as already remarked, two solutions of the equations were made, termed A and B.

Solution A is that derived from the meridian observations alone. Solution B is that of the normal equations formed from both the meridian observations and the transits.

The results of the solutions in the case of Mercury are shown in the following tables. The relation of the unknown quantities given in the first columns, A and B, to the corrections of the elements has been shown in a preceding section (§ 27). The upper half of the table shows the corrections to the elements; the lower half those of the secular variations.

It will be seen that all the values, with a single exception, come out less than a unit. In stating the corrections to the elements, it must be remembered that, owing to the proximity of Mercury to the Sun, the errors of geocentric place are much less than those of the heliocentric elements, so that an error in the latter indicates a proportionally smaller error in the actual observations. For the same reason we must expect a less degree of precision in the elements as finally derived than in the case of the other planets.

5690 N ALM——5

MERCURY.

Results of solutions of the normal equations.

	Unknowns.		Factors.	Corrections of elements.		
Symbol.	A.	B.		Symbol.	A.	B.
$[m']$	—0.1478	—0.1207	0.1	$\delta m : m$	—0.0148 ''	—0.0121 ''
l	—0.1342	—0.0752	4.	δl	—0.537	—0.301
J	—0.2436	—0.2299	3.	δJ	—0.731	—0.690
N	—0.0227	—0.0201	3.	$\text{Sin } J\, \delta N$	—0.068	—0.061
ε	+0.2074	+0.2194	1.	$\delta \varepsilon$	+0.207	+0.219
e	—0.1202	+0.4094	3.	δe	—0.361	+1.228
π	+0.5209	+0.2688	10.	$\delta \pi$	+5.209	+2.689
e''	+0.0669	+0.8397	0.6	$\delta e''$	+0.040	+0.504
π''	—0.2248	—0.7027	0.6	$e'' \delta \pi''$	—0.135	—0.422
r''	+1.1240	+1.0566	2.	$\delta r''$	+2.248	+2.113
δ	—0.2310	—0.2556	1.	δ	—0.231	—0.256
l''	—0.0354	—0.0897	1.	$\delta l''$	—0.035	—0.090
a	+0.4803	+0.4930	1.	a	+0.480	+0.493
l'	—0.2060	—0.1209	16.	$D_t \delta l$	—3.296	—1.935
J'	—0.0114	+0.0636	12.	$D_t \delta J$	—0.137	+0.764
N'	+0.1000	+0.0930	12.	$\text{Sin } J\, D_t \delta N$	+1.200	+1.116
ε'	+0.0681	+0.0966	4.	$D_t \delta \varepsilon$	+0.272	+0.386
e'	—0.1165	+0.0987	12.	$D_t \delta e$	—1.398	+1.184
π'	—0.2385	—0.0252	40.	$D_t \delta \pi$	—9.540	—1.008
$e''{}'$	—0.1968	+0.1317	2.4	$D_t \delta e''$	—0.472	+0.316
$\pi''{}'$	—0.1677	—0.1193	2.4	$e'' D_t \delta \pi''$	—0.402	—0.286
$r''{}'$	+0.1108	+0.0806	4.	$D_t \delta r''$	+0.886	+0.645
δ'	—0.1826	—0.1233	4.	$D_t \delta$	—0.730	—0.493
$l''{}'$	—0.1442	—0.3152	4.	$D_t \delta l''$	—0.577	—1.261
a'	—0.3160	—0.1973	4.	$D_t a$	—1.264	—0.789

Mean epoch of corrections, 1865.0.

Discordance in the observed Right Ascensions of Mercury.

33. The most remarkable feature in the result is the value of the quantity represented by $[r'']$. The unknown quantity introduced with this symbol had as its coefficient the derivative of the geocentric place as to the Earth's radius vector, and the result would therefore be an apparent constant correction to that radius vector. Since, however, the position of the planet depends only on the ratio of the distances of the Earth and Mercury, it follows that the actual correction may be regarded as a correction to the ratio of the mean distances.

The determination of the mean distances by KEPLER's third law may be regarded as so unquestionable that the true value of this unknown quantity should be regarded as zero, and the result as a purely fictitious one, arising from erroneous elements of reduction or systematic personal errors. It was the possibility of the latter that led to its introduction. When the planet is east of the Sun, observations are always made on or near its west limb, or at least on some point west of the true center, and *vice versa*. The value of $\delta r''$ therefore indicates that there is a remarkable systematic difference in the observed Right Ascension according as the planet is east or west of the Sun, and therefore according to the illuminated side. The sign of the result shows that the reduction to the center of the planet was apparently too small. It is therefore of interest to learn according to what law this error changed as the planet moved around its relative orbit.

It has up to the present time been impracticable to substitute the unknown quantities in the original equations of condition, and thus determine the separate residuals, and for the purpose of investigating the present case such a substitution is the less necessary, owing to the smallness of the unknown quantities. I have therefore simply determined the mean correction to the Right Ascension given by all the observations during the various periods in six segments of the relative orbit, near the elongations, and before and after the two conjunctions. The results are shown in the following table. Commencing with the moment of inferior conjunction, column A contains the mean correction to the tabular Right Ascension, from observations made within about twenty days following. Column B contains the observations made from twenty days after the inferior conjunction until twenty days before superior conjunction, a period during which the planet was generally near its greatest west elongation. Column C contains the observations made during the twenty days following and up to superior conjunction. Then follow in regular order the corresponding results when the planet was east of the Sun, beginning with the twenty days following superior conjunction and going around to inferior conjunction.

Table showing the mean corrections to the tabular Right Ascension of Mercury in six segments of its relative orbit.

Epochs.	A		B		C	
	"	*wt.*	*"*	*wt.*	*"*	*wt.*
1765–1791	+3.24	4	+2.61	5	------------	
1793–1815	+2.06	6	+1.82	10	+0.97	4
1817–1839	+3.06	6	+1.79	24	+1.13	24
1840–1849	+1.46	6	+1.48	18	—0.38	20
1850–1859	+3.72	4	+0.77	20	+0.08	16
1860–1869	+1.18	28	+1.14	72	+0.31	44
1870–1880	+1.18	25	+0.74	65	—0.20	61
1881–1892	+1.19	38	+0.98	63	—0.15	62

Epochs.	D		E		F	
	"	*wt.*	*"*	*wt.*	*"*	*wt.*
1765–1791	+0.92	1.5	+1.30	10	+0.81	3
1793–1815	+2.82	5	+1.10	16	+1.85	5
1817–1839	+0.27	25	+3.76	24	—1.29	5
1840–1849	+0.22	22	—0.53	30	+0.75	3
1850–1859	+0.69	14	—0.39	28	—0.65	4
1860–1869	—0.44	55	—0.55	69	—0.35	16
1870–1880	—0.52	57	—1.25	67	—0.30	24
1881–1892	—0.84	80	—0.73	102	—0.37	26

The remarkable feature of these results is the near approach to constancy in the values of the numbers in each column, after the secular variation is allowed for, and the large magnitude of the corrections. The most natural conclusion is that the reduction from the limb of the planet or the observed center of light to the true center was too small by an amount which, at the mean distance of the Sun, must have been nearly or quite a second of arc (*cf.* § 3). The adopted semidiameter 3″.4 seems so well established, both by micrometric measures and by heliometer measures during transits of Mercury, that such a correction to the diameter seems inadmissible.

I have not yet been able to enter upon the investigation of the source of this anomaly. A very important question is that of its influence on the results. Since a constant error in the radius vector of a planet would have opposite effects on the elements in different points of the relative orbit, it may be inferred that the effect of the error would be nearly eliminated in an extensive series of observations distributed equally between the two elongations. Actually, however, there seems to have been an appreciable lack of symmetry in this respect, as the influence of the unknown quantity upon the other unknowns is not inconsiderable. Although the law of change, as shown in the preceding table, does not correspond to the magnitude of the coefficient of $\delta r''$, this coefficient being relatively too small near inferior conjunction and too large near superior conjunction, it is still probable that through the introduction and elimination of $\delta r''$ a large part of the injurious effect is eliminated.

Comparison of transits and meridian observations of Mercury.

34. Another remarkable result which may be associated with this is shown by the difference between the solutions A and B, in the case of the eccentricity and perihelion not only of the planet, but of the Sun. It will be seen that the meridian observations alone give a negative correction to the eccentricity of the planet, while, when the transits are included, the correction becomes positive. That this is due to a systematic cause running through the observations is shown by the fact that the same thing is true of the secular variation of the eccentricity. This relation of the correction to its secular variations holds true for three of the four relative elements, and for the eccentricity and perihelion both of the planet and of the Earth. In the case of the Earth's perihelion, however, there is a nearer approach to conformity between the two results.

There is yet another anomaly in this connection, which indi-cates a very considerable systematic error in the older meridian observations, which is not completely eliminated from the ele-ments. If we take the values of the unknown quantities and their secular variations, which result from the two solutions, and substitute them in the linear functions of the corrections to the elements derived from the transits alone, namely

$$V = 1.487\,\delta l - 0.487\,\delta\pi - 1.137\,\delta e - 1.01\,\delta l'' + 1.19\,e''\delta\pi''$$
$$+ 1.58\,\delta e''$$
$$W = 0.716\,\delta l + 0.284\,\delta\pi + 0.896\,\delta e - 0.97\,\delta l'' - 1.11\,e''\delta\pi''$$
$$- 1.62\,\delta e''$$

we find the following results:

From meridian observations $V = -2''.99 + 0''.69\,T$
From November transits $-1\ .69 - 2\ .63\,T$
From combined solution $-2\ .77 - 2\ .30\,T$

From meridian observations $W = +0''.89 - 4''.55\,T$
From May transits alone $+1\ .39 + 1\ .84\,T$
From combined solution $+1\ .39 + 0\ .42\,T$

We conclude that, had no transits ever been observed, the errors of the elements and their secular variations, derived from the great mass of meridian observations, would have caused an error of some $5''$ per century in the heliocentric place of the planet at the times of the May transits, and of some $3''$ at the time of the November transits.

The fact that the combined solution B satisfies the transits so much better than A, although the total weight of equations A is so much greater than that of the transit equations, shows that the meridian observations give only weak results for the functions in question.

Meridian observations of Venus.

35. So far as the meridian observations are concerned, those of Venus were treated on the same general plan as the observations of Mercury. The following are the principal points of difference:

1. The hypothetical quantity $\delta r''$ is omitted. Hence no index to the consistency of the observations at different points of the relative orbit can be derived from the solution.

2. Tenths of a unit were included in the coefficients of the equations, and no modification was made in the units. The units and tenths were, however, dropped in the final solution of the normal equations.

Results of observed transits of Venus.

36. We put, at the time of a transit,

v, the longitude in orbit of Venus;
l, its mean longitude, or the mean value of v;
β, λ, its ecliptic latitude and longitude;
L, the Sun's true longitude.

Then

$$\delta\lambda = \cos\ i\ \delta v + \sin^2 i\ \delta\theta$$
$$= 0.9982\ \delta v + 0.0592 \sin i\ \delta\theta$$

We thus have, for the dates of the observed transits,

$$1761\text{-}'69;\quad \delta\beta = -\ 0.0592\ \delta v + 0.9982 \sin i\ \delta\theta$$
$$1874\text{-}'82;\quad \delta\beta = +\ 0.0592\ \delta v - 0.9982 \sin i\ \delta\theta$$

I have discussed very fully the observations of the transits of 1761 and 1769 in *Astronomical Papers*, Vol. II. The final results which I shall use are found on page 404 of that volume. Here I have put.

$$x,\ \text{correction to}\ \lambda - \mathrm{L};$$
$$-\ y,\ \text{correction to}\ \beta,$$

the Sun's latitude being supposed to require no correction. The values of x and y for 1769 are distinguished by an accent. I have also represented by z_2 and z_3 the corrections to the difference of the semidiameters of the Sun and planet, for the respective internal contacts, to which may be added the unknown but probably nearly constant quantity due to personal error in estimating the time of contact. From their very nature these quantities do not admit of accurate determination, and must therefore be eliminated from the equations. From the observations of internal contact are derived the following four equations:

$$1761\quad \mathrm{II};\ -.87\,x\ +.50\,y\ +z_2 = -\ 0''.07$$
$$\mathrm{III};\ +.68\quad +.73\quad +z_3 = -\ 0''.06$$
$$1769\quad \mathrm{II};\ -.64\,x' -.77\,y' + z_2 = -\ 0''.27$$
$$\mathrm{III};\ +.84\quad -.55\quad +z_3 = +\ 0''.02$$

We have here more unknown quantities than equations, so that it is not practicable to determine them all separately. What I have done has been first to assume $z_2 = z_3$. This presupposes that the distance of centers at the estimated apparent contact at egress is, in the general mean, the same as at ingress. The result of any error in this hypothesis will be almost completely eliminated from the mean latitude at the two transits, but not from the longitude.

Still, the values of x and y can not be separately determined; I have therefore so combined the equations as to obtain mean values of x and y for the two contacts, assuming that this would be the result of supposing these quantities to have the same values at both epochs. Calling these values x'' and y'', we have by addition and subtraction, supposing $z_2 = z_3$,

$$- 0.39\,x'' + 2.55\,y'' = 0''.12$$
$$3.03\,x'' + 0.45\,y'' = 0''.30$$

We thus have*

$$x'' = + 0''.09$$
$$y'' = + 0''.06$$

These corrections are not applicable to the coordinates from LEVERRIER'S tables as they stand, but to those quantities as corrected by the following amounts:

$$\Delta \lambda = + 0''.25$$
$$\Delta \beta = + 2''.00$$

---

* In a second approximation to these quantities, which may be made after the correction to the centennial motion of the node is determined, we should put, on account of this correction,

$$y = y'' - 0''.11$$
$$y' = y'' + 0''.11$$

The solution would then give

$$y'' = + 0'.06$$
$$x'' = + 0''.14$$

I have carried through a more careful approximation in a subsequent chapter.

We thus find, for the corrections to LEVERRIER'S tables at the epoch 1765.5,

$$\delta \lambda - \delta L = + 0''.09 + 0''.25 = + 0''.34$$
$$\delta \beta \quad = - 0''.06 + 2''.00 = + 1''.94$$

and hence

$$\delta v = + 0''.22 + 0''.998 \, \delta L$$
$$\sin i \, \delta \theta = + 1''.95 + 0''.059 \, \delta L$$

A still farther modification is required to the tabular longitude on account of the correction to the mass of the Earth used by LEVERRIER, and hence to the periodic perturbations in longitude. This correction is $+ 0''.20$. We thus have for the correction to the orbit longitude of Venus—

$$\delta v = + 0''.02 + 0''.998 \, \delta L$$

For the results of the transits of 1874 and 1882 I have depended entirely on the heliometer measures and photographs made by the German and American expeditions, respectively. The definitive results of the German observations, as worked up by Dr. AUWERS, are found in Vol. V of the German Reports on the Transits.* The American photographic measures of 1874 have not been officially worked up and published, but a preliminary investigation from the data contained in the published measures was made by D. P. TODD, and published in the *American Journal of Science*, Vol. 21, 1881, page 491. The measures of 1882 have been definitively worked up by HARKNESS, but only the results published. They are found in the report of the Superintendent of the U. S. Naval Observatory for the year 1890.

The corrections to the geocentric Right Ascension and Declination of Venus relative to the Sun thus derived are

---

* Die Venus-durchgänge 1874 und 1882 Bericht über die Deutschen Beobachtungen Fünfter Band, Berlin, 1893.

given in the following table. In taking the mean the weights are not strictly those which would result from the probable errors as assigned, but, in accordance with a general principle, independent results have received a weight more near to equality than would be indicated by the mean errors.

$$1874: \text{German}, \quad \delta \text{ R. A.} = + 4.77 \pm 0.28$$

$$\text{American}, \quad . \quad . \quad . \quad + 4.14 \pm 0.30$$

$$\text{Adopted}, \quad . \quad . \quad . \quad + 4.44$$

$$\text{German}, \quad \delta \text{ Dec.} = + 2.28 \pm 0.10$$

$$\text{American}, \quad . \quad . \quad . \quad + 2.50 \pm 0.30$$

$$\text{Adopted}, \quad . \quad . \quad . \quad + 2.34$$

$$1882: \text{German}, \quad \delta \text{ R. A.} = + 9.03 \pm 0.12$$

$$\text{American}, \quad . \quad . \quad + 9.10 \pm 0.08$$

$$\text{Adopted}, \quad . \quad . \quad . \quad + 9.07$$

$$\text{German}, \quad \delta \text{ Dec.} = + 2.02 \pm 0.06$$

$$\text{American}, \quad . \quad . \quad . \quad + 2.02 \pm 0.08$$

$$\text{Adopted}, \quad . \quad . \quad . \quad + 2.02$$

We change these results successively to geocentric longitude and latitude, heliocentric longitude and latitude, and orbital longitude and latitude. The results of these several changes are as follow:

	1874.	1882.
Corr. in geoc. long.	$+ 3''.853$	$+ 8''.077$
Corr. in lat.	$+ 2 .724$	$+ 2. 971$
Corr. in hel. long.	$- 1 .415$	$- 2 .965$
Corr. in hel. lat.	$+ 1 .001$	$+ 1 .091$
Corr. in orbital long.	$- 1 .35$	$- 2 .90$
Value of sin $i \delta \theta$	$- 1 .08$	$- 1 .26$

Equations from transits of Venus.

37. The corrections to the heliocentric positions of Venus and the Earth, as thus found, are now to be expressed in terms of corrections to the elements. The results of this expression are shown in the following equations:

Equations given by the corrections to the orbital longitude.

I. *Epoch,* 1765.5; $\tau = -3.90$; weight $= 200$

$$0.992 \, \delta l + 1.17 \, e\delta\pi + 1.62 \, \delta e - 0.976 \, \delta l'' - 1.81 \, e''\delta\pi'' - 0.85 \, \delta e''$$
$$= +0''.02 \pm 0.''15$$

II. *Epoch,* 1874.9; $\tau = +0.48$; weight $= 400$

$$-0''.88\mu + 1.009 \, \delta l - 1.223 \, e\delta\pi - 1.596 \, \delta e - 1.030 \, \delta l''$$
$$+1.864 \, e''\delta\pi'' + 0.817 \, \delta e'' = -1''.35 \pm 0''.08$$

III. *Epoch,* 1882.9; $\tau = +0.80$; weight $= 800$

$$0''.60\mu + 1.008 \, \delta l - 1.146 \, e\delta\pi - 1.651 \, \delta e - 1.028 \, \delta l'' + 1.825 \, e''\delta n''$$
$$+ 0.900 \, \delta e'' = -2''.90 \pm 0.''027$$

Equations given by the corrections to the orbital latitude.

I. 1765.5; $\sin i\delta\theta - 0.057 \, \delta l'' - 0.11 \, e''\delta\pi'' - 0.05 \, \delta e'' = +1''.95$
$$\pm 0''.10$$

II. 1874.9; $\sin i\delta\theta - 0.061 \, \delta l'' + 0.110 \, e''\delta\pi'' + 0.048 \, \delta e'' = -1''.08$
$$\pm 0''.04$$

III. 1882.9; $\sin i\delta\theta - 0.061 \, \delta l'' + 0.107 \, e''\delta\pi'' + 0.053 \, \delta e'' = -1''.26$
$$\pm 0''.019$$

The weights assigned to these three equations are, respectively, 200, 600, and 1,600.

Before using these equations the corrections to the elements were transformed into the unknown quantities defined in § 27, and their secular variations by multiplying the coefficients by the factors given on page 56.

Solutions of the equations for Venus.

38. The parts of the normal equations formed from the preceding conditional equations were added to the parts from the meridian observations, and the resulting solution B obtained. As in the case of Mercury, a solution A was made of the normal equations derived from the meridian observations alone. The results are as follows:

VENUS.

Results of solutions of the normal equations.

Unknowns.			Factors.	Corrections of elements.		
Symbol.	A.	B.		Symbol.	A.	B.
[m]	—0.0708	—0.0834	7.	$\delta m : m$ $''$	—0.496 $''$	—0.584 $''$
[l]	—0.1435	—0.1501	5.	δl	—0.718	—0.751
[J]	+0.1156	+0.1340	6.	δJ	+0.694	+0.804
[N]	+0.0164	+0.0106	7.	$\sin J \, \delta N$	+0.115	+0.074
[e]	+0.0941	+0.1003	3.	δe	+0.282	+0.301
[π]	+0.0628	+0.0764	3.	$e \, \delta \pi$	+0.188	+0.229
[ε]	+0.0246	+0.0271	4.	$\delta \varepsilon$	+0.098	+0.108
[e'']	+0.0336	+0.0318	2.5	$\delta e''$	+0.084	+0.080
[π'']	—0.0274	—0.0212	2.	$e'' \delta \pi''$	—0.055	—0.042
[a]	+0.4742	+0.4642	1.	a	+0.474	+0.464
[δ]	—0.0383	—0.0375	5.	δ	—0.192	—0.188
[l'']	—0.0768	—0.0743	4.	$\delta l''$	—0.307	·—0.297
[l]$'$	—0.1846	—0.1983	20.	$D_t \delta l$	—3.692	—3.966
[J]$'$	+0.0970	+0.1088	24.	$D_t J$	+2.328	+2.611
[N]$'$	—0.0561	—0.0594	28.	$\sin J \, D_t N$	—1.571	—1.663
[e]$'$	+0.1472	+0.1644	12.	$D_t e$	+1.766	+1.973
[π]$'$	+0.0555	+0.0698	12.	$e \, D_t \pi$	+0.666	+0.838
[ε]$'$	+0.0182	+0.0202	16.	$D_t \varepsilon$	+0.291	+0.323
[e'']$'$	+0.0283	+0.0317	10.	$D_t e''$	+0.283	+0.317
[π'']$'$	+0.0399	+0.0506	8.	$e'' D_t \pi''$	+0.319	+0.405
[a]$'$	—0.0820	—0.0347	4.	$D_t a$	—0.328	—0.139
[δ]$'$	—0.0020	—0.0002	20.	$D_t \delta$	—0.040	—0.004
[l'']$'$	·—0.0562	—0.0662	16.	$D_t \delta l''$	—0.899	—1.059

Mean epoch of correction, 1863.0

Comparison of transits of Venus with meridian observations.

39. To show to what extent the results of the meridian observations differ from those of the observed transits over the Sun, we form the values of the absolute terms of the equations of condition, §37, first by substituting the values A of the corrections, and then the values B. We thus have:

Residuals in orbital longitude.

	1765.5.	1874.9.	1882.9.
(α) From meridian obs. alone .	$-0''.07$	$-1''.36$	$-2''.54$
(β) From combined solution .	$+0''.04$	$-1''.43$	$-2''.78$
(γ) From transits alone . . .	$+0''.02$	$-1''.35$	$-2''.90$
Discordance, $(\gamma)-(\alpha)$ . .	$+0''.09$	$+0''.01$	$-0''.36$
Discordance, $(\gamma)-(\beta)$ . .	$-0''.04$	$+0.08$	$-0''.12$

Residuals in orbital latitude.

	1765.5.	1874.9.	1882.9.
(α) From meridian obs. alone .	$+1''.92$	$-0''.77$	$-0''.96$
(β) From combined solution .	$+2''.06$	$-0''.91$	$-1''.12$
(γ) From transits alone . . .	$+1''.95$	$-1''.08$	$-1''.26$
Discordance, $(\gamma)-(\alpha)$ . .	$+0''.03$	$-0''.31$	$-0''.30$
Discordance, $(\gamma)-(\beta)$ . .	$-0''.11$	$-0''.17$	$-0''.14$

It will be seen that the combined solution represents the observations of the transits much better here than in the case of Mercury.

Solution of the equations for Mars.

40. As the formation of the normal equations for Mars was approaching its end, a singular discordance among the residuals of the partial normal equations for different periods was noticed. On tracing the matter out it appeared that while the correction of the geocentric longitude of LEVERRIER's tables in 1845 and again in 1892 was quite small, the correction in 1862 was considerable. Now there is an inequality of long period, about forty years, in the mean motion of Mars, depending on the action of the Earth, and having for its argument $15g'-8g$. This coefficient is of the seventh order in the eccentricities, and the terms of the ninth or even of the eleventh order might be sensible in a development in powers of the eccentricities and sines or cosines of multiples of the mean longitudes. The conclusion which I reached was that the theoretical value of this coefficient was not determined with sufficient precision. As the work of solving the equations could not wait for a new determination and a new formation of the absolute terms of the normal equations, it was decided to make an approximate empirical correction to the theory. This was used to correct the absolute terms of the partial normal equa-

tions for each period, and the solution was then proceeded with. The chances seem to be that by this process the injurious effect of the error upon the elements derived from the equations would be inconsiderable; this is, however, a point on which it is impossible to speak with certainty. It is the intention of the writer to recompute the doubtful terms of the perturbations, and, if possible, reconstruct the absolute terms of the normal equations in accordance with the corrected theory. Meanwhile, the present work necessarily rests on the imperfect theory with the approximate empirical corrections, which are as follow:

$$\delta l = 0''.30 \cos (15g' - 8g - 223°)$$
$$e\delta\pi = 0''.15 \cos (15g' - 8g)$$

As the elements of Mars are derived wholly from meridian observations, only one set of equations of condition was formed. The results of the solution are shown in the following table:

MARS.

Unknowns.		Factors.	Corrections of elements.	
Symbol.	Value.		Symbol.	Value.
$[m']$	$-.02278$	0. 3	$\delta m : m$	-0.007
				$''$
$[l]$	$-.44854$	2.	δl	-0.897
$[J]$	$+.05479$	2. 5	δJ	$+0.137$
N	$+.06724$	2. 5	Sin $J \delta N$	$+0.168$
ϵ	$+.43803$	$\frac{10}{1}$	$\delta \epsilon$	$+0.626$
π	$-.05056$	$\frac{100}{1}$	$\delta \pi$	-0.722
e	$+.07474$	4.	$\delta \varepsilon$	$+0.299$
ϵ''	$-.49898$	2.	$\delta \epsilon''$	-0.998
π''	$-.42409$	2.	$\epsilon'' \delta \pi''$	-0.848
a	$+.18545$	5.	a	$+0.927$
δ	$-.04536$	5.	δ	-0.227
$[l'']$	$+.05786$	3.	$\delta l''$	$+0.174$
$[l]'$	$+.16605$	8.	$D_t \delta l$	$+1.328$
$[J]'$	$+.13408$	10.	$D_t J$	$+1.341$
$N]'$	$-.02263$	10.	Sin $J D_t N$	-0.226
$\epsilon]'$	$-.03180$	$\frac{10}{1}$	$D_t \epsilon$	-0.182
$\pi]'$	$-.00928$	$\frac{100}{1}$	$D_t \pi$	-0.530
$e]'$	$+.06097$	16.	$D_t \varepsilon$	$+0.976$
$[\epsilon'']'$	$-.12597$	8.	$D_t \epsilon''$	-1.008
$\pi'']'$	$+.00853$	8.	$\epsilon'' D_t \pi''$	$+0.068$
$a]'$	$-.09670$	20.	$D_t a$	-1.934
$\delta]'$	$-.01168$	20.	$D_t \delta$	-0.234
$[l'']'$	$+.13111$	12.	$D_t \delta l''$	$+1.573$

Reference to the ecliptic.

41. In all the preceding determinations the planes of the orbits are referred to the plane of the Earth's equator, or, to speak more exactly, to a plane through the Sun parallel to the Earth's equator. As in astronomical practice the ecliptic is taken as the fundamental plane, it is necessary to investigate the reduction of the elements from one plane to the other.

Let us consider the spherical triangle formed on the celestial sphere by the plane of the orbit, the plane of the ecliptic, and the plane of the Earth's equator. For the sides and opposite angles of this triangle we have

$$
\begin{array}{llll}
\text{Sides:} & N & \theta & \psi \\
\text{Opposite angles:} & i & 180^\circ - J & \varepsilon
\end{array}
$$

When equatorial coordinates are used, the position of the planet is considered as a function of the three quantities

$$N; \quad J; \quad \varepsilon \tag{a}$$

When ecliptic coordinates are used, the three corresponding quantities are

$$\theta; \quad i; \quad \varepsilon \tag{b}$$

Taking the set of quantities (a) as the fundamental parts of the triangle, and expressing the corrections of the other parts as functions of them, we have

$$
\begin{aligned}
\delta i &= + \cos \psi \delta J + \sin \psi \sin J \delta N - \cos \theta \delta \varepsilon \\
\sin i \, \delta \theta &= - \sin \psi \delta J + \cos \psi \sin J \delta N + \cos i \sin \theta \delta \varepsilon
\end{aligned} \tag{c}
$$

Taking (b) as the fundamental parts, we have for the corrections to N and J

$$
\begin{aligned}
\delta J &= \cos \psi \delta i - \sin \psi \sin i \delta \theta + \cos N \delta \varepsilon \\
\sin J \, \delta N &= \sin \psi \delta i + \cos \psi \sin i \delta \theta - \cos J \sin N \delta \varepsilon
\end{aligned} \tag{d}
$$

The numerical values assigned to the coefficients in these equations are those corresponding to the mean epoch 1850. The fact that they change somewhat in the course of a hundred years has not been taken account of. The future astronomer will meet with a real difficulty in that the corrections to a

set of elements at one epoch do not accurately correspond to
similar corrections at another epoch. It is impossible to do
away rigorously with the difficulty thus arising, except by
introducing a more general system of elements than elliptic
ones. The error is, happily, not important in the present state
of astronomy. The equations in question for the three planets
are as follow:

Mercury.

$$\delta i = + .799 \, \delta J + .602 \sin J \, \delta N - .688 \, \delta \varepsilon$$
$$\sin i \delta \theta = - .002 \, \delta J + .799 \sin J \, \delta N + .721 \, \delta \varepsilon$$

Venus.

$$\delta i = + .373 \, \delta J + .928 \sin J \, \delta N - .255 \, \delta \varepsilon$$
$$\sin i \delta \theta = - .928 \, \delta J + .373 \sin J \, \delta N + .967 \, \delta \varepsilon$$

Mars.

$$\delta i = \quad .703 \, \delta J + .712 \sin J \, \delta N - .664 \, \delta \varepsilon$$
$$\sin i \delta \theta = - .712 \, \delta J + .703 \sin J \, \delta N + .747 \, \delta \varepsilon$$

For the inverse relations we have—

Mercury.

$$\delta J = .799 \, \delta i - .602 \sin i \delta \theta + .983 \, \delta \varepsilon$$
$$\sin J \, \delta N = .002 \, \delta i + .799 \sin i \delta \theta - .162 \, \delta \varepsilon$$

Venus.

$$\delta J = .373 \, \delta i - .928 \sin i \delta \theta + .990 \, \delta \varepsilon$$
$$\sin J \, \delta N = .928 \, \delta i + .373 \sin i \delta \theta - .125 \, \delta \varepsilon$$

Mars.

$$\delta J = .703 \, \delta i - .712 \sin i \delta \theta + .998 \, \delta \varepsilon$$
$$\sin J \, \delta N = .712 \, \delta i + .703 \sin i \delta \theta - .052 \, \delta \varepsilon$$

CHAPTER IV.

COMBINATION OF THE PRECEDING RESULTS TO OBTAIN THE MOST PROBABLE VALUES OF THE ELEMENTS AND OF THEIR SECULAR VARIATIONS FROM OBSERVATIONS ALONE.

In the two preceding chapters are derived four separate values of the six corrections, α, δ, $\delta\varepsilon$, $\delta l''$, $\delta e''$, and $e''\delta\pi''$, and of their secular variations, which pertain to the orbit and motion of the Earth relative to the stars. We have now to combine these four results so as to derive the most probable values of the twelve unknown quantities in question.

Deviations from the method of least squares.

42. If we applied without modification the principles of the method of least squares, we should first eliminate the elements and secular variations for each planet from the normal equations given by observations of that planet, which would leave us with three sets of normal equations, containing only the twelve quantities depending on the motion of the Earth. We should then reduce these normal equations to equality of weight, by multiplying each of them by the appropriate factors, and we should then consider the observed corrections to the solar elements derived from observations of the Sun alone as affording equations of condition to be reduced to the adopted system of weights, and then multiplied by their coefficients and added to the normal equations. The solution of the single set of normal equations thus formed would lead to the definitive values of the solar elements and of their secular variations, which, being substituted in the eliminating equations from each planet, would lead to the definitive elements of the planet and of their secular variations.

This proceeding is not, however, advisable in the present case, because, owing to the immense mass of material worked

5690 N ALM——6

up, the errors to be principally feared are not the accidental
ones, of which alone the method of least squares takes account,
but the systematic ones arising principally from personal
equation and imperfect reduction of the observations to the
actual center of the planet or of the Sun. These errors affect
different elements in very different ways and to different
amounts; from some they will be almost completely eliminated and from others they will not. We must therefore proceed by a tentative process, ascertaining at each step, so far
as possible, how each result will come out before we accept it
as final, to be combined with other results. In doing this it
is necessary to deviate so widely from what are commonly
regarded as fundamental principles of the theory of the combination of observations that a brief presentation of the principles involved is appropriate.

It is frequently accepted as an axiom that when we have
several non-accordant determinations of the same quantity,
between which we have no reason for choosing, the most probable value is the arithmetical mean. The operation of taking
the arithmetical mean is, in fact, the simplest application of
the method of least squares. The fundamental hypothesis on
which this method rests is that the probability of an error of
magnitude $\pm\, x$ is given by the well-known exponential equation

$$\varphi\,(h,\ x)\ dx = \frac{h}{\sqrt{\pi}}\, e^{-h^2 x^2}\, dx \qquad\qquad (a)$$

h, the modulus of precision, being a constant. It was shown by
GAUSS that this function for the probability follows rigorously
from the principle of the arithmetical mean. It therefore follows that the method of the arithmetical mean, and therefore
that of least squares, is rigorously correct only so far as the
law of error is expressed by the above exponential function.

It scarcely needs to be pointed out that, as a matter of fact,
the law of error in question is not true. Not only so, but in
astronomical experience it deviates from the truth in a way
admitting of precise statement. It presupposes that the modulus of precision is a determinate quantity. Were this the
case, then, to take a single instance, the probability of an

error five times as great as the probable error would be less than 0.001, and the probability of an error six times as great would be about 0.0001. This is not true, because, taking the function $\varphi(h, x)$ as a basis, we may say that the modulus of precision, h, is nearly always in practice an uncertain quantity. Let us then put

$$h_1, \ h_2, \ h_3, \quad . \quad . \quad .$$

for the possible values of h, and

$$p_1, \ p_2, \ p_3, \quad . \quad . \quad .$$

for the several probabilities that h has these respective values. Then the probability function will become

$$\varphi(x) = p_1 \, \varphi(h_1, x) + p_2 \, \varphi(h_2, x) + \quad . \quad . \quad . \qquad (b)$$

Now this form can not be reduced to the form (a) with any value whatever of the modulus h. If we make the closest representation possible, we shall have a curve in which small values and large values of x are relatively less probable as compared with the facts than are intermediate values. To show that this is the actual case, let us suppose that we have three determinations of an unknown quantity. If we proceed in the usual way, we should infer the value of h, the measure of precision, from the discordance of these three values. But it is evident that this determination of h would be very uncertain. Should the three values chance to be fortunately accordant, then, proceeding in the usual way, our function would lead to the conclusion that the probability of an error of a certain magnitude in the mean was very small, when, as a matter of fact, it might be very considerable.* The value of h being

---

* To take a simple and quite possible instance, let three observations of a star with a meridian circle give, for the seconds of declination, $0''.4, 0''.5$, and $0''.6$. By the canons of least squares the mean result would be

$$0''.50 \pm 0''.039$$

and the probability of an error as great as $0''.1$ would come out about 0.08.

uncertain, the true form of the function is not (*a*) but (*b*). It follows that we may lay down the following general rule:

The best value from a system of non-accordant determinations is not the arithmetical mean, but a mean in which less weight is assigned to those results which deviate most widely from the mean of the others.

I have considered the subject from this point of view in the *American Journal of Mathematics*, Vol. VIII, p. 343, and given tables for determining the weights to be assigned to the results when the law of error is that derived from several hundred observed contacts of the limb of Mercury with that of the Sun during transits of the planet.

Another well-known defect in the method of least squares is that it does not take any account of systematic errors. The greater the number of observations that are combined, the larger the proportion in which the errors of the results may be due to the systematic errors in the observations or the elements of reduction. Although such errors may elude investigation so far as their determination and elimination is concerned, we may yet be able to point out their origin, and to show to what extent they would influence each separate result. Of some results we can say with entire confidence that they are but slightly affected with systematic error; of others, that they may be very largely so affected. In the latter case, the weights of the results, as determined from the solution of the normal equations, give no clue whatever to the probable magnitude of the error.

The result of this is that in the following paper we are more than once confronted with the following problem: Among several determinations of a quantity one is known to be free from systematic error and to be affected with a well determined probable mean error, $\pm \varepsilon$. There are also one or more other determinations of which the probable error is unknown and can not be determined, because we have no sufficient knowledge of the probable effect of systematic errors upon the result. What shall be the relative weight assigned to two such results in order to obtain the mean? The decision of this question is necessarily a matter of judgment, the grounds for which it might be extremely prolix to state at length. An attempt has been made in these cases to classify the results, so as to give a general idea of what is likely to be their modulus of precision, and weight them accordingly.

Any attempt at numerical accuracy in such an estimate would be labor thrown away. It has therefore been considered sufficient in such cases to state what the conclusion of the author is, leaving its revision and criticism to the future investigator. Indeed, in some cases, as in that of the correction to the centennial motion of the Sun in longitude, a convenient round number has been chosen, very near to the result of well-determined weight.

We should be carrying the preceding conclusions too far if they led us to a general distrust of the conclusions reached by the method of least squares. The doctrines that there is a necessary limit to the accuracy with which astronomical determinations can be made; that systematic errors necessarily affect every such determination; and that the canons of least squares necessarily lead to illusory probable errors, are too sweeping. We may lay down the general rule that if we have a sufficient number of really independent determinations of an unknown quantity, of which we individually know nothing except that they are the results of actual measures, and not mere guesses, then the arithmetical mean will be a definite result, the probable deviation from which will actually follow the law given by the canons in question with a closeness which will continually increase with the number of independent determinations.

If we have such knowledge of the relative values of the various determinations as to assign greater weight to some than to others, the result will be still better when those weights are used, provided always that they are assigned without undue bias in favor of those results which most nearly approach the value supposed to be approximately correct.

These considerations lead me to a policy which I have always adopted when it was easy to do so in the following discussions, namely, that of so conducting the work as to lead to as many independent determinations of a quantity as possible, and of always giving a less relative weight to such sets of determinations as might from any cause whatever be supposed affected by an important common source of error. Where the independent determinations are few in number, the computation of a definite probable error is impracticable, and the probable mean error assigned is necessarily a result of a judgment based on all the circumstances.

Relative precision of the two methods of determining the elements of the Earth's orbit.

43. When the system of determining the solar elements from observations of the planets as well as of the Sun was originally decided upon, it was supposed that the two methods would give results not greatly differing in accuracy in the case of any of the elements. This, however, is proved by the results not to be the case. Attention has already been called to the extreme consistency of the values found for the correction to the eccentricity and perihelion of the Earth's orbit from observations of the Sun. This consistency inspires us with confidence that the probable errors of the corrections to the elements as given do not exceed a few hundredths of a second. But the determination of these elements from observations of Mercury and Venus may be seriously affected by the form of the visible disks of those planets, which results in observations being made only upon one limb when east of the Sun and the other limb when west of it. Thus personal equation and the uncertainty of the semidiameter to be applied in each case may have an effect upon the result. But personal equation is likely to be smaller in the case of Mercury than in that of Venus, owing to the smallness of its disk.

There is another circumstance which weakens the independent determination of the Earth's eccentricity and perigee from observations of the planets. If we define the orbit of a planet, not as a curve, but as the totality of points which the planet occupies at a great number of given equidistant moments during its revolution, then it is easy to see that the general mean effect of an increase of the eccentricity is to displace the entire orbit toward the point of the celestial sphere marked by its aphelion, while the effect of a change of its perihelion is to move the entire orbit in its own plane in a direction at right angles to the line of apsides. The result is that in a series of observations of a planet from the Earth the corrections to the eccentricity and perihelia of the two orbits can not be entirely independent, and we can determine with entire precision only two linear functions expressive of the relative displacements just described. It may be admitted that, were observations exactly similar in kind made around the entire relative orbit in equal numbers, the effect of the principle systematic errors would be nearly eliminated from the result. But we can not rely upon this being the case, and even were it the case there would probably be a residual effect which would be large in proportion to the interdependence of the two sets of correc-tions. But in this connection the important remark is to be made that, so far as these systematic errors are invariable, they would not affect the secular variations, but only the abso-lute values of the elements. We may therefore assign greater relative weights to the former than to the latter.

So far as we can classify the results, I have concluded that in the case of the secular variations of ε, e'', and π'', the weight of the determination from Mercury and Venus might receive a weight one-fifth that from the Sun. But in the case of the absolute values of these quantities, it would seem from the discordance of the results that the relative weight of the planetary results should be much smaller.

In dealing with the common error, α, of the adopted Right Ascensions of the stars, it is to be remarked that we may regard the observations in Right Ascension as fitted to give the values of $\alpha + \delta l''$, while $\delta l''$ necessarily depends solely upon the observations of declination, in effect if not in form. Hence, although the unknown quantities of the solution are α and $\delta l''$, I have deemed it best to derive the result by regarding $\alpha + \delta l''$ as the quantity to be first found, instead of α itself.

Secular variations of the solar elements.

44. The following table shows the corrections to the tabular secular variations of the solar elements, as they have been found from observations. In the cases of Mercury and Venus the results of both solutions are given for the sake of compari-son, although only solution B is used. The relative weights

have been determined by the considerations already set forth. In the case of Mars, the final determinant of the solution for the solar elements came out so nearly evanescent as to show that no reliable values could be obtained, a result which we

Corrections to the secular variations of the solar elements derived from observations only.

	$D_t \delta \varepsilon$		$D_t \delta l''$		$D_t \delta e''$	
	"	*w.*	*"*	*w.*	*"*	*w.*
From observations of—						
The Sun	+0.48	5	—0.97	1	+0.23	5
Mercury, solution A	+0.27		—0.58		—0.47	
" " B	+0.39	1	—1.26	1	+0.32	1
Venus, solution A	+0.29		—0.90		+0.28	
" " B	+0.32	1	—1.06	1	+0.32	1
Mars	+1.03	½				
Mean	+0.48		—1.10		+0.26	
Adopted	+0.48		—1.00		+0.21	

	$e'' D_t \delta \pi''$		$D_t (a + \delta l'')$		$D_t a$
	"	*w.*	*"*	*w.*	*"*
From observations of—					
The Sun	+0.32	5	—0.63	2	+0.34
Mercury, solution A	—0.40		—1.84		—1.26
" " B	—0.29	1	—2.05	3	—0.79
Venus, solution A	+0.32				—0.33
" " B	+0.46	1	—1.20	2	—0.14
Mars					
Mean	+0.25		—1.40		—0.30
Adopted	+0.26		—1.30		—0.30

might expect, because, in order to separate the principal elements of the Earth's orbit from those of the planet, observations should be continued all around the relative orbit, whereas, as a matter of fact, they are generally made only near the time of opposition. I have judged, however, that the correction to the secular variation of the obliquity obtained by putting $D_t \delta l'' = -1''.00$ in the equation for $D_t \delta \varepsilon$ might enter with half the weight that it does in the cases of Mercury and Venus. Before the final values and weights of the quantities in the table had been definitively revised, provisional values were used in the subsequent part of the investigation.

These provisional values are given in the last line of the table. It is also to be noted that the secular variations of e, e'' π, π'' and ε in the definitive theory and tables are those computed from the adopted masses of the planets.

Correction to the standard of Declination.

45. The results for the secular variation of δ, the common error of the standard Declinations within the zodiacal limits, are not given in the table, as other data are available for its determination. The following shows the separate values of δ and its secular variation, derived from observations of the planets to Saturn inclusive. For reasons already stated observations of the Sun are not used for this purpose.

			$Wt.$
From observations of Mercury,	$\delta =$	$- 0.18 - 0.49\,T\,;$	2
Venus,		$- 0.19 - 0.00\,T\,;$	1
Mars,		$- 0.21 - 0.23\,T\,;$	4
Jupiter,		$- 0.04 - 0.43\,T\,;$	3
Saturn,		$+ 0.04 - 0.68\,T\,;$	4

$$\text{Mean;} \quad \delta = - 0''.09 - 0''.42\,T$$
$$\text{Adopted;}\; \delta = - 0\ .08 - 0\ .50\,T$$

Not only observations of the planets but those of the fixed stars are available for the determination of δ and of its secular variation. In the discussion of the Declinations derived from observations with the Greenwich and Washington transit circles (*Astronomical Papers*, Vol. II), I have shown that the Greenwich observations indicate, with some uncertainty, a secular variation of the corrections to the standard declinations which will give a value of about $-0''.55$ for the secular variation of δ. But BRADLEY'S Declinations, as reduced by AUWERS, would give a still larger negative value, approximating to an entire second. As the value which we may assume for δ does not greatly influence the other elements, I have adopted as a convenient probable result, the variation $-0''.50\,T$.

Definitive secular variations of the planetary elements from observations alone.

46. Having decided upon the adopted values of the six quantities found in the last article, we regard them as known quantities, and substitute them in the eliminating equations, which give the values of the remaining secular variations. As the unknown quantities in these equations are not the corrections themselves, but certain functions of them, we prepare the following table, showing the formation of the quantities which are to be substituted in the several equations. The table scarcely seems to need any explanation, except that the unknown quantities given in the three columns on the right are formed by dividing the secular variations for twenty-five years by the coefficients given in § 27.

Adopted secular variations of the solar unknowns, to be substituted in the eliminating equations for the several planets.

		Mercury.	Venus.	Mars.
$D_t \, \delta l''$	$= -1''.00;$	$[\, l'' \,]' = -0.250;$	$-0.0625;$	$-0.0833;$
$D_t \, \delta$	$= -0 \;.50;$	$[\, \delta \,]' = -0.125;$	$-0.0250;$	$-0.0250;$
$D_t \, \alpha$	$= -0 \;.30;$	$[\, \alpha \,]' = -0.075;$	$-0.0750;$	$-0.0150;$
$e'' D_t \, \delta \pi''$	$= +0 \;.26;$	$[\, \pi'' \,]' = +0.108;$	$+0.0325;$	$+0.0325;$
$D_t \, \delta e''$	$= +0 \;.21;$	$[\, e'' \,]' = +0.087;$	$+0.0210;$	$+0.0262;$
$D_t \, \delta \varepsilon$	$= +0 \;.48;$	$[\, \varepsilon \,]' = +0.120;$	$+0.0300;$	$+0.0300.$

To facilitate a judgment or rediscussion of this part of the process, we give on the next three pages the normal equations between all the secular variations which remain after the corrections to the elements of the Sun and planets are eliminated from the original normal equations. We give these rather than the eliminating equations which were actually used in the substitution, because they show more fully the relations between the unknown quantities, and can therefore be better used in any ulterior discussion. Regarding the preceding six quantities as known, and substituting them in the normal equations for the secular variations, we derive the definitive values of the secular variations which relate to the planets. They are shown in the next table. In the latter the values of the solar elements are repeated for the sake of completeness.

MERCURY.

Matrix of normal equations for the secular variations after the elements are eliminated.

[The equations are to be completed by adding the terms to complete a symmetric matrix.]

$[l]'$	$[J]'$	$[N]'$	$[e]'$	$[\pi]'$	$[\varepsilon]'$	$[e'']'$	$[\pi'']'$	$[r'']'$	$[l'']'$	$[\alpha]'$	$[\delta]'$	n
+50736	+1358	−2332	−7035	−17877	−52	+1412	+3499	−240	−5654	+3967	−262	= −5733
	+20868	−13243	+505	−5916	−69	+85	+2186	+503	+609	−3360		= −69
		+26189	−461	+181	+6197	−760	−306	−241	+375	+928	+2829	= +1924
			+32240	+21741	−1492	−15000	+1487	−1804	+3693	+3390	−7343	= +105
				+34227	−1147	−9316	−10018	+492	−841	−2053	−3044	= +4434
					+9409	+394	−189	−673	−284	+310	+5675	= +346
						+10061	+421	+1213	−753	−1118	+4127	= −135
							+9024	−1181	+4537	+3966	−1200	= −3247
								+5321	−1228	−1306	+73	= +1273
									+13442	+10407	−1062	= −5859
										+11218	+14	= −6237
											+18195	= −1253

VENUS.

Matrix of normal equations for the secular variations after the elements are eliminated.

[The equations are to be completed by adding the terms to complete a symmetric matrix.]

	$[U]'$	$[J]'$	$[N]'$	$[e]'$	$[\pi]'$	$[\varepsilon]'$	$[e'']'$	$[\pi'']'$	$[\alpha]'$	$[\delta]'$	$[V']'$	n
$[U]'$	+10462	− 287	− 368	+ 6413	+3307	+ 195	− 3611	−4646	−173	− 161	− 9041 =	− 498
$[J]'$		+22521	− 5758	− 50	− 274	−15945	+ 65	+ 549	+ 14	− 640	+ 1152 =	+2452
$[N]'$			+17333	+ 176	− 71	+ 5116	− 378	− 209	− 88	+ 822	+ 118 =	−1467
$[e]'$				+11545	+4245	− 67	− 8727	−6911	−194	− 1947	+ 5852 =	+ 671
$[\pi]'$					+8432	+ 142	+ 1498	−7684	−132	+ 955	− 3651 =	+ 395
$[\varepsilon]'$						+14585	− 6	− 361	− 15	+ 1984	+ 831 =	−1746
$[e'']'$							+10491	+1928	+137	+ 2593	+ 3404 =	− 383
$[\pi'']'$								+8717	+234	− 393	+ 4683 =	− 501
$[\alpha]'$									+296	0	+ 1326 =	− 82
$[\delta]'$										+10707	− 29 =	− 240
$[V']'$											+13139 =	+ 122

MARS.

Matrix of normal equations for the secular variations after the elements are eliminated.

[The equations are to be completed by adding the terms to complete a symmetric matrix.]

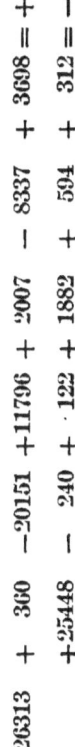
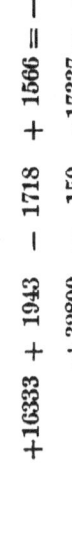

	$[I]'$	$[J]'$	$[N]'$	$[e]'$	$[\pi]'$	$[\varepsilon]'$	$[e']'$	$[\pi'']'$	$[\alpha]'$	$[\delta]'$	$[V_1]'$	n
$[I]'$	$+26138$	-12	-621	$+1150$	-5153	-923	$+3784$	-1908	$+24171$	-1354	-21784	$= -1363$
$[J]'$		$+23064$	$+307$	$+321$	-724	-21582	$+728$	-139	-1278	$+3714$	$+341$	$= +1796$
$[N]'$			$+29197$	$+792$	-1037	$+4753$	$+1181$	-89	$+8483$	$+7793$	$+51$	$= -1503$
$[e]'$				$+23444$	-3202	$+117$	$+13957$	$+14180$	$+1726$	$+4954$	-208	$= -2381$
$[\pi]'$					$+26313$	$+360$	-20151	$+11796$	$+2007$	-8337	$+3698$	$= +1977$
$[\varepsilon]'$						$+25448$	-240	$+122$	$+1882$	$+594$	$+312$	$= -1725$
$[e']'$							$+22363$	-1236	-1119	$+7167$	-2436	$= -2697$
$[\pi'']'$								$+16333$	$+1943$	-1718	$+1566$	$= -553$
$[\alpha]'$									$+32800$	$+150$	-17337	$= -1593$
$[\delta]'$										$+41909$	$+1351$	$= -1162$
$[V_1]'$											$+20224$	$= +1051$

Values of the secular variations as derived from observations only.

		Unknown.	Corr.	Tables.	Result.
			$''$	$''$	$''$
Mercury.	$D_t e$	$-.0691$	-0.83	$+ 4.19$	$+ 3.36 \pm 0.50$
	$e D_t \pi$	$+.1577$	$+1.30$	$+116.94$	$+118.24 \pm 0.40$
	$D_t i$	$+.0593\,J$	$+0.83$	$+ 6.31$	$+ 7.14 \pm 0.80$
	$\sin i\,D_t\,\theta$	$+.0815\,N$	$+0.70$	$- 92.59$	$- 91.89 \pm 0.50$
	$D_t\,\delta l$	$-.0967$	-1.55		
Venus.	$D_t e$	$+.1393$	$+1.67$	$- 11.13$	$- 9.46 \pm 0.20$
	$e D_t \pi$	$+.0685$	$+0.82$	$- 0.53$	$+ 0.29 \pm 0.20$
	$D_t i$	$+.1153\,J$	-0.65	$+ 4.52$	$+ 3.87 \pm 0.30$
	$\sin i\,D_t\,\theta$	$-.0592\,N$	-2.73	-102.67	-105.40 ± 0.12
	$D_t\,\delta l$	$-.1919$	-3.84		
Earth.	$D_t e$		$+0.21$	$- 8.76$	$- 8.55 \pm 0.09$
	$e D_t \pi$		$+0.26$	$+ 19.22$	$+ 19.48 \pm 0.12$
	$D_t \varepsilon$		$+0.48$	$- 47.59$	$- 47.11 \pm 0.25$
Mars.	$D_t e$	$-.1190$	-0.68	$+ 19.68$	$+ 19.00 \pm 0.27$
	$e D_t \pi$	$+.0536$	$+0.29$	$+149.26$	$+149.55 \pm 0.35$
	$D_t i$	$+.1136\,J$	$+0.17$	$- 2.43$	$- 2.26 \pm 0.20$
	$\sin i\,D_t\,\theta$	$-.0442\,N$	-0.76	$- 71.84$	$- 72.60 \pm 0.20$
	$D_t\,\delta l$	$-.0946$	-0.76		

The first column of numbers in this table gives the unknown quantity as found immediately from the eliminating equations. These quantities being multiplied by the factors given in § 27, we have the corrections to the tabular secular variations, as given in the column "correction." The next column gives the value of the tabular secular variations, which are in all cases those actually adopted by LEVERRIER. In the case of the Earth, as has been pointed out by STÜRMER and by INNÈS, the secular variation of the radius vector does not correspond to that of the longitude. But as that of the longitude is the preponderating quantity in its effect on geocentric places, I have regarded the value of the eccentr.city used in the tables of the equation of the center as the tabular one to be adopted.

The numbers in the column "Unknown," which are followed by the letters J and N, are the respective values of $\lfloor J \rfloor_i$ and $\lfloor N \rfloor_i$, which are changed to δi and sin $i \, \delta \, \theta$ by the equations of § 41.

Finally, we have the results given in the last column for the actual secular variations of the several elements as derived from the preceding discussion of all the observations.

The result is followed by the probable mean error of each of the quantities as estimated from the probable magnitude of the sources of error to which they are liable. As in other cases, these quantities are very largely a matter of judgment, because the probable errors as determined in the usual way from the eliminating equations would be entirely unreliable.

Definitive corrections to the solar elements for 1850.

47. Leaving the above results to be subsequently discussed, we go on with the solution of the equations. By a continuation of the process just described, we regard the preceding secular variations as known quantities, and substitute them in the eliminating equations for the solar elements which are derived from the normal equations for each planet. By this substitution, we reach three fresh sets of values of the corrections of the solar elements themselves, one set from the observations of each planet, which are to be reduced to 1850 and combined with those already found from observations of the Sun, in order to obtain the most probable result.

Here we meet with the same difficulty that confronted us in the case of the secular variations. With the exception of the Sun's mean longitude, we are to regard the results derived from each of the planets as affected by obscure sources of systematic error, the probable magnitude of which can only be inferred from the general deviation of the quantities themselves. As in the former case, α is not regarded as a quantity independently determined, but $\alpha + \delta l''$ has been taken instead. The concluded value of α is then found by subtracting $\delta l'$, from $\delta l'' + \alpha$. Since the corrections to the solar elements pertain to each separate epoch, those derived from the observations of the planets are severally reduced to 1850, and the results are shown in the following table:

Separate values of the corrections to the solar elements for 1850, after the above definitive values of the secular variations are substituted in the eliminating equations from solution B, reduced to 1850.

	$\delta\epsilon$	$\delta l''$	$\delta e''$	$e''\delta\pi''$	$a+\delta l''$	a
From observations of—	″	″	″	″	″	″
The Sun	−.30	+.05	+.10	.00	−.02	−.07
Mercury	+.13	+.07	+.48	−.47	+.60	+.53
Venus	+.13	−.17	+.06	−.07	+.34	+.50
Mars	+.25	+.24	−.83	−.82	+1.18	+.94
Adopted	−.20	−.02	+.12	−.04	+.46	+.48

These adopted values are employed in the subsequent stages of the discussion, but are not in all cases regarded as definitive. In the case of ϵ the value $-0''.20$ is that which I have actually used in the subsequent determinations of the elements, but for the final value of the obliquity it will be seen that I have taken $-0''.15$ as more probable.

CHAPTER V.

MASSES OF THE PLANETS DERIVED BY METHODS INDE-PENDENT OF THE SECULAR VARIATIONS WITH THE RESULTING COMPUTED SECULAR VARIATIONS.

48. The plan of discussion laid down in Chapter I contemplates the determination of the masses of each of the planets from all data independent of the secular variations, in order to determine how far the observed secular variations can be reconciled with these masses. The following is a summary of these determinations. The planets outside of Jupiter need no discussion, as the well-known determinations of their masses are amply accurate for all our present purposes.

Mass of Jupiter.

49. One of the works connected with the present subject has been the determination of the mass of Jupiter from the motions of(33), Polyhymnia. My work on this subject has not yet been printed in full, but I have given in *Astronomische Nachrichten*, No. 3249 (Bd. 136, S. 130), a brief summary of the results. The mass of Jupiter has been derived not only from the motions of Polyhymnia, but from such other sources as seemed best adapted to give a reliable result. The following table, transcribed from the publication in question, shows the separate results and the conclusions finally reached:

Reciprocal of mass of Jupiter from—

		Wt.
All observations of the satellites,	1047.82	1
Action on FAYE's comet (MÖLLER),	1047.79	1
Action on Themis (KRUEGER),	1047.54	5
Action on Saturn (HILL),	1047.38	7
Action on Polyhymnia,	1047.34	20
Action on WINNECKE's comet (V. HAERDTL),	1047.17	10

$$1047.35$$
$$\text{m. e. } \pm 0.065$$

It will be seen that the result from observations of the satellites has been assigned a very small weight. This course has been indicated by the circumstances. Other conditions being equal, the greater the mass of a planet the less the proportionate precision with which that mass can be determined by observations on the satellites. In any case, if the measures of the distances between the satellites and the primary are in error by a small fraction, a, of their whole amount, then the error of the mass will be in error by the fraction $3a$ of its amount. For reasons founded on the construction and use of the heliometer, I doubt whether the absolute measures made with those forms of that instrument which have been used in determining the mass of Jupiter can be relied upon within their three-thousandth part. If so, the determination of the mass of the planet itself would be doubtful by its thousandth part in each separate case. The chance of personal equation between transits of the satellites and the planet vitiates in the same way the results from observed transits of the planet and satellites. Notwithstanding the great refinement of the discussion by KEMPF of observations made at Potsdam, and the care with which he, SCHUR, and others have determined the mass of Jupiter by a discussion of all the observations of the satellites, I can not conceive that the probable error of any possible result they could derive would be less than 0.3 or 0.4 in the denominator.

In this connection the discordances between the mass of Saturn, found by Prof. HALL and by other observers from observations of the satellites, are worthy of consideration. They lead us to suspect that perhaps it is through good fortune rather than by virtue of their absolute reliability that determinations of the mass of Jupiter from observations of the satellites have agreed so well.

As to the weights assigned to the other results, only the last needs especial mention. The probable error assigned by v. HAERDTL to his result is very much smaller than that which I find for the mean of all the results. But, as remarked in the paper in question, it has received a smaller relative weight than that corresponding to its assigned probable error, because of distrust on my part whether observations on a comet can be considered as having always been made on the center of gravity of a well-defined mass, moving as if that center were a material point subject to the gravitation of the Sun and planets. This distrust seems to be amply justified by our general experience of the failure of comets to move in exact accordance with their ephemerides.

I propose to accept the value thus found,

$$\text{Mass of Jupiter} = 1 \div 1047.35$$

as the definitive one to be used in the planetary theories.

Mass of Mars.

50. In consequence of the minuteness of the mass of Mars, measures of its satellites, especially the outer one, afford a value of its mass much better than can be derived by its action on the planets. When nearest the earth, the major axis of the orbit of the outer satellite subtends an angle of 70″. I can not think that the systematic error to be feared in the best measures, such as those made by Prof. HALL, can be as great as half a second. It therefore appears to me that the mean error in adopting Prof. HALL's value of the mass does not exceed its fiftieth part. This is a degree of precision much higher than that of any determination through the action of Mars on another planet.

Prof. HALL's measures of 1892 show a minute increase of the mean distance given by his work of 1877. The result is—

$$\nu''' = +\, 0.014$$

These observations, however, were made when the position of the orbit of the satellite was unfavorable to an exact determination of the elements of motion. I have adhered to the original value in the work of the present chapter.

Mass of the Earth.

51. I have already pointed out the difficulty in the way of determining the mass of the Earth from its action on the other planets. On the other hand, the solar parallax has, in recent years, been determined in various ways with such precision that the mass of the Earth to be used in the plan-

etary theories can best be derived from it. The theory of the
relation between the mass of the Earth and its distance from
the Sun, as given by observations of the seconds pendulum
and the length of the sidereal year, is one of the best established results of celestial mechanics. It is, in effect, the
principle on which the lunar theory is constructed. In this
theory the disturbing action of the Sun is necessarily a function of the ratio of the mass of the Sun to that of the Earth.
But in the accepted theory this ratio is eliminated through
the ratio of the lunar month to the sidereal year. From the
well-established ratio between the distance of the Moon and
the length of the seconds pendulum, the ratio of the masses
of the Sun and Earth come out of this theory with great
precision. It need not be developed here; it will suffice to
give the numerical result, which is that between the ratio M
of the mass of the Sun to that of the Earth and the mean
equatorial horizontal parallax of the Sun in seconds of arc
there exists the relation

$$\pi^3 M = [8.35493]$$

I have derived seven values of the solar parallax by different
methods, of which the following are the preliminary results:

	$''$	$''$	Wt.
GILL's observations of Mars, 1877,	8.780	± .020	1
Contact observations, transits of Venus.	8.794	± .018	1
Aberration and velocity of light,	8.798	± .005	16
Parallactic equation of the Moon,	8.799	± .007	5
Measures of small planets on GILL's plan,	8.807	± .007	8
LEVERRIER's method,	8.818	± .030	0.5
Measures of Venus from Sun's center,	8.857	± .022	1

Mean result, $\pi = 8''.802 \pm 0''.005$

I have provisionally taken this mean as the most probable
value of the solar parallax derived from all sources except the
mass of the Earth. The above relation then gives

$$M = 332\,040$$

Taking for the mass of the Moon $1 \div 81.52$, we have for the ratio of the combined masses of the Earth and Moon to the mass of the Sun

$$m'' = \frac{1}{328\,016}$$

a result of which the probable error may be regarded as something more than a thousandth part of its whole amount.

Mass of Venus.

52. The mass of Venus adopted in the provisional theory, to which LEVERRIER'S tables were reduced, was $.000\,002\,4885$ $= 1 \div 401847$, which is that of LEVERRIER'S tables of Mercury. In the preceding discussions the following three factors of correction to this mass have been found:

From observations of the Sun . . $-.0118 \pm .0034$
From observations of Mercury . . $-.0121 \pm .0050$
From observations of Mars . . . $-.0076 \pm .(?)$
Mean . . . . $-.0119 \pm .0028$

The mean error assigned to the result from observations of the Sun may be regarded as real, because the result is the mean of a great number of completely independent determinations, among which no common error is either *a priori* probable or shown by the discordance of the results. In the case of Mercury, however, as already remarked, the effect of systematic errors is such that, although they are almost completely eliminated from the result, the mean error computed in the usual way would be misleading. The weight assigned is therefore largely a matter of judgment.

The fact that it was necessary to introduce an empirical correction, with a period of about forty years, into the mean longitude of Mars, vitiates the determination of the mass of Venus from its action on that planet, because one of the principal terms in the action of Venus on Mars has a period which does not differ from forty years enough to make the determination of the mass independent of this empirical correction. I have therefore assigned no weight to the result. We thus have for the mass of Venus, as derived from the periodic perturbations of Mercury and the Earth produced by its action.

$$m' = 1 \div 406\,690 \pm 1140$$

Mass of Mercury.

53. The mass of Mercury which I have heretofore adopted, $1 \div 7\,500\,000$, was rather a result of general estimate than of exact computation. The fact is that the determinations of this mass have been so discordant, and varied so much with the method of discussion adopted, that it is scarcely possible to fix upon any definite number as expressive of the mass. An examination of LEVERRIER'S tables of Venus shows that with the mass of Mercury there adopted ($1 : 3\,000\,000$) Mercury frequently produces a perturbation of more than one second in the heliocentric longitude of Venus. When the latter is near inferior conjunction, the actual perturbation will be more than doubled in the geocentric place, so that the latter might not infrequently be changed by $1''$, even if the mass of Mercury be less than one-half LEVERRIER'S value. It was therefore to be expected that a fairly reliable value of the mass of Mercury would be obtained from the periodic perturbations of Venus.

Referring to § 27, it will be seen that the indeterminate mass of Mercury appears in the equations in the form

$$\frac{1 + 7\mu}{3\,000\,000}$$

From the solution B, § 38, the value of μ comes out

$$\mu = -0.0834$$

corresponding to a mass of Mercury of $1 : 7\,210\,000$. But in a subsequent solution of the equations, when the secular variations are determined from theory and substituted in the normal equation for μ, we find

$$\mu = -0.0889$$

which gives

$$m = 1 \div 7\,943\,000$$

The work of the present chapter is based on the former value.

A consideration of the probable error of this result is important. The fortuitous errors which mostly affect it are of the
class which I have termed *semi-systematic*. Under this term I
include that large class of errors which, extending through or
injuriously affecting a limited series of observations, cause the
probable error of a result to be larger than that given by the
solution of the equations, but which, nevertheless, like purely
accidental ones, would be eliminated from the mean result of
an infinite series of observations. To this class belong the
errors arising from personal equation in observing the limb of
Venus, or, what is the same thing, a difference between the
practical semidiameter corresponding to the observer and that
adopted in the reductions. We may suppose that, during a
period of several days, when Venus is not far from inferior
conjunction, its geocentric position is affected by a perturbation produced by Mercury. Through the error alluded to, all
the observations made by any one observer, and in fact all
that are made anywhere, may be affected by a certain constant error in Right Ascension. Near another inferior conjunction the same state of things may be repeated, with the
perturbation in the opposite direction. If, now, the observations were made by the same observer, and under the same
circumstances, the personal error would be eliminated from
the mean of these two results so far as the mass of Mercury is
concerned. But very frequently different observers will have
made the observations under the two circumstances, and different conditions will have prevailed. Thus, it is only through
the general law of averages that we can expect the effect of
these fortuitous but systematic errors to be completely eliminated. That they would be eliminated in the long run is
evident from the fact that there can be no permanent relation between the personal equations of the observers and the
changes in the action of Mercury upon Venus. Moreover,
Venus has been observed with a fair degree of accuracy
through more than half a century, and it seems reasonable
to suppose that during that time the errors in question would
nearly disappear.

It is clear from these considerations that the probable
error derived from the solution of the equations would be

entirely misleading. But a probable error which ought to be reliable can be obtained by a process similar to that which I have adopted elsewhere in this paper, namely, dividing up the materials into periods, and determining the probable error from the discordances among the results of the several periods. This probable error will be reliable, because there is no reason why the same error should affect the mass of Mercury through any two periods. I therefore take the partial normal equations in μ derived from Right Ascensions during the several periods, substitute in them the values of the unknown quantities found from solution B, μ excepted, and thus form sixteen partial normal equations in μ. These equations may be changed into the corresponding equations of condition, of weight unity, by dividing each by the square root of the coefficient of the unknown quantity. The residuals then left when the definitive value of the unknown quantity is substituted will be those from whose discordance the probable error may be inferred.

The partial normal equations thus found from the Right Ascensions are as follow:

1750–'62.	44 μ =	− 38		1830–'40.	5649 μ =	− 831
1765–'74.	1265	− 165		1840–'49.	2913	− 18
1775–'86.	15	− 5		1849–'56.	2238	− 49
1787–'96.	209	+ 53		1857–'64.	4506	− 129
1796–'06.	345	+ 19		1865–'71.	7736	− 265
1806–'14.	439	+ 135		1871–'79.	7062	− 761
1814–'19.	942	+ 2		1879–'86.	4958	− 407
1820–'30.	1786	− 330		1885–'92.	9561	− 1306

$$\text{Sum:}\quad 49\,668\ \mu = -\ 4095$$
$$\mu = -\ 0.0824 \pm .019$$

The difference between this value of μ, which is obtained only to find the probable error, and that formerly found, arises principally from the omission of the declination equations. The mean error corresponding to weight unity comes out

$$\varepsilon_1 = \pm\ 4''.2$$

which, as anticipated, is much larger than that which would
be given by the discordance of the original observations.
This does not mean that the original observations are affected
by any such mean error as $\pm\ 4''.2$, but that the discordances
between the 16 values of μ are as great as we should expect
them to be if the original observations were absolutely free
from systematic error, but affected by purely accidental errors
of this mean amount.

The results of the solution for the mass of Mercury may be
expressed in the form

$$m = \frac{1 \pm 0.32}{7\ 210\ 000} \quad \text{and} \quad \frac{1 \pm 0.35}{7\ 943\ 000}$$

In all researches which have been made on the motion of
ENCKE'S comet by ENCKE, VON ASTEN, and BACKLUND, the
determination of this mass has been kept in view. The
results are, however, so discordant that, as already remarked,
scarcely any definitive result can be derived from them.

To this statement there is, however, one apparent exception.
In an appendix to his very careful and elaborate discussion of
WINNECKE'S comet, VON HAERDTL has derived the value of
the mass of Mercury from all the return of ENCKE'S comet as
worked up by VON ASTEN and BACKLUND.* The only inter-
pretation which I can put upon his result is this: If we regard
the acceleration of the comet, which it is supposed results
from all the observations made upon it, as non-existent, the
following two masses of Mercury are derivable from the observations:

$$1819\text{--}1868,\ m = 1 \div 5\ 648\ 600 \pm 2000$$
$$1871\text{--}1885,\ m = 1 \div 5\ 669\ 700 \pm 600\ 000$$

He also finds, from the motion of WINNECKE'S comet,

$$m = 1 \div 5\ 012\ 842 \pm 697\ 863$$

---

* Denkschriften der Kaiserlichen Akademie der Wissenschaften, Vol.
56, p. 172–175. Vienna, 1889.

and from four equations of LEVERRIER

$$1 \div 5\,514\,700 \pm 100\,000$$

The consistency of these results seems to me entirely beyond what the observations are capable of giving, and I hesitate to ascribe great weight to them. Moreover, the result implicitly contained in these numbers, that the supposed secular acceleration of the comet disappears when we attribute the preceding mass to Mercury, merits farther inquiry.

The probable density of the planet may form a basis for at least a rude estimate of its probable mass. The fact that the Earth, Venus, and Mars have densities not very different from each other, while that of the Moon is 0.6 the density of the Earth, leads us to suppose that Mercury, being nearest to the Moon in mass, has probably a slightly greater density. Its diameter at distance unity has been repeatedly measured and found to be 6".6, or, roughly speaking, three-eighths that of the Earth. Were its density 0.7, its mass would therefore be about 1 : 9,000,000. In view of the fact that the measured diameter is probably somewhat too small, these considerations lead us to conclude that the mass is probably between 1 : 6,000,000 and 1 : 9,000,000.

As the value of the mass to be used in investigating the secular variations, I have adopted

$$\nu = + 0.08$$

$$\text{Mass of Mercury} = \frac{1.08}{7\,500\,000}$$

Secular variations resulting from theory.

54. In the *Astronomical Papers*, Vol. V, Part IV, were computed the secular variations of the elements of the orbits in question using, as the basis of the work, the values of the

masses whose reciprocals are found in the column A below.
In column B are cited the masses which I have decided upon.

	A Original reciprocal of mass.	B Adpoted reciprocal of mass.	ν
Mercury,	7 500 000	6 944 444	$+.080$
Venus,	410 000	406 750	$+.0080$
Earth + Moon,	327 000	328 000	$-.00305$
Mars,	3 093 500	3 093 500	0
Jupiter,	1047.88	1047.35	$+.00050$
Saturn,	3501.6		0
Uranus,	22 756		0
Neptune,	19 540		0

In the case of the Earth we have to add the secular variation of the perihelion produced by the non-sphericity of the
system Earth + Moon. For the principal term I have found,

$$D_t\, e''\, \delta\, \pi'' = +\, 0''.129$$

The resulting values of the secular variations, expressed as
functions of ν, ν', ν'', ν''', are given in the following section:

Theoretical secular variations for 1850.

Mercury.

$$
\begin{array}{llllllll}
D_t\, e & =+ & 4.22 & +0.00\nu+ & 2.8\nu'+ & 1.1\nu''- & 0.1\nu''' & =+ & 4.24 \\
e\,D_t\, \pi_1 & =+109.36 & +0.00 & +56.8 & +18.8 & +0.5 & & =+109.76 \\
D_t\, i & =+ & 6.76 & -0.04 & - 0.6 & - 1.4 & +0.0 & =+ & 6.76 \\
\sin i\, D_t\, \theta & =- & 92.12 & -0.33 & -49.3 & -12.2 & -1.2 & =- & 92.50
\end{array}
$$

Venus.

$$
\begin{array}{llllllll}
D_t\, e & =- & 9.58 & -1.30\nu+ & 0.0\nu'- & 4.9\nu''- & 0.2\nu''' & =- & 9.67 \\
e\,D_t\, \pi_1 & =+ & 0.39 & -0.81 & + 0.0 & - 3.9 & +0.5 & =+ & 0.34 \\
D_t\, i & =+ & 3.43 & +0.76 & + 0.0 & + 0.0 & -0.3 & =+ & 3.49 \\
\sin i\, D_t\, \theta & =-105.92 & +0.26 & -29.2 & -43.2 & -1.2 & & =-106.00
\end{array}
$$

Earth.

$$D_t e \quad = - \quad 8.57 \; -0.12\nu + \; 1.3\nu' \qquad\qquad -1.6\nu''' = - \quad 8.57$$
$$e\,D_t\,\pi \quad = + \; 19.36 \; -0.18 \; + \; 5.8 \qquad\qquad +1.6 \quad = + \; 19.39$$
$$D_t\,\varepsilon \quad = - \; 46.65 \; -0.21 \; -28.3 \qquad\qquad -0.7 \quad = - \; 46.89$$

Mars.

$$D_t e \quad = + \; 18.71 \; +0.03\nu + \; 0.1\nu' + \; 2.1\nu'' \qquad\qquad = + \; 18.71$$
$$e\,D_t\,\pi_1 \quad = +148.82 \; +0.06 \; + \; 4.6 \; +21.4 \qquad\qquad = +148.80$$
$$D_t\,i \quad = - \quad 2.34 \; -0.04 \; +12.0 \; + \; 0.0 \; +0.0\nu''' = - \quad 2.25$$
$$\sin i\,D_t\,\theta = - \; 72.43 \; -0.27 \; -25.1 \; - \; 7.4 \; -1.0 \quad = - \; 72.63$$

CHAPTER VI.

EXAMINATION OF THE HYPOTHESES BY WHICH THE DEVIATIONS OF THE SECULAR VARIATIONS FROM THEIR THEORETICAL VALUES MAY BE EXPLAINED.

55. The investigations of the present chapter are founded on a comparison of the secular variations derived purely from observations in Chapter IV, with those resulting from the values of the masses obtained independently of the secular variations in the last chapter. For the sake of clearness, these two sets of secular variations and their differences are collected in the following table. The mean errors assigned to the theoretical values are those which result from the probable mean errors of the respective masses. They are therefore not to be regarded as independent. The mean errors given in the column of differences are those which result from a combination of those of the other two columns. The errors of the observed quantities must not, however, be judged from those of the differences, because subsequent changes in the masses of Mercury, Venus, and the Earth may produce a general diminution in the discordances.

Mercury.

	Observation.	Theory.	Diff.	$\varDelta$	$\sqrt{w}.$
$D_t\,e$	$+\ \ \ 3.36 \pm 0.50$	$+\ \ \ 4.24 \pm .01$	$-0.88 \pm .50$	-0.86	2
$e\,D_t\,\pi$	$+118.24 \pm 0.40$	$+109.76 \pm .16$	$+8.48 \pm .43$	$.\ \ .$	0
$D_t\,i$	$+\ \ \ 7.14 \pm 0.80$	$+\ \ \ 6.76 \pm .01$	$+0.38 \pm .80$	$+0.38$	$1\frac{1}{4}$
$\sin i\,D_t\,\theta$	$-\ 91.89 \pm 0.45$	$-\ 92.50 \pm .16$	$+0.61 \pm .52$	$+0.23$	2.2

Venus.

	Observation.	Theory.	Diff.	$\varDelta$	$\sqrt{w}.$
$D_t\,e$	$-\ \ \ 0.46 \pm 0.20$	$-\ \ \ 9.67 \pm .24$	$+0.21 \pm .31$	$+0.12$	5
$e\,D_t\,\pi$	$+\ \ \ 0.29 \pm 0.20$	$+\ \ \ 0.34 \pm .15$	$-0.05 \pm .25$	$.\ \ .$	0
$D_t\,i$	$+\ \ \ 3.87 \pm 0.30$	$+\ \ \ 3.49 \pm .14$	$+0.38 \pm .33$	$+0.44$	$3\frac{1}{5}$
$\sin i\,D_t\,\theta$	-105.40 ± 0.12	$-106.00 \pm .12$	$+0.60 \pm .17$	$+0.52$	8

109

Earth.

	Observation.	Theory.	Diff.	Δ	$\sqrt{w}$.
$D_t e$	$-\ \ 8.55\pm0.09$	$-\ \ 8.57\pm.04$	$+0.02\pm.10$	$+0.02$	10
$e\,D_t\,\pi$	$+\ 19.48\pm0.12$	$+\ 19.38\pm.05$	$+0.10\pm.13$	. .	.
$D_t\,\varepsilon$	$-\ 47.11\pm0.23$	$-\ 46.89\pm.09$	$-0.22\pm.27$	-0.46	$4\tfrac{1}{3}$

Mars.

$D_t e$	$+\ 19.00\pm0.27$	$+\ 18.71\pm.01$	$+0.29\pm.27$	$+0.29$	3.7
$e\,D_t\,\pi$	$+149.55\pm0.35$	$+148.80\pm.04$	$+0.75\pm.35$	. .	0
$D_t\,i$	$-\ \ 2.26\pm0.20$	$-\ \ 2.25\pm.04$	$-0.01\pm.20$	$+0.08$	5
$\sin i\,D_t\,\theta$	$-\ 72.60\pm0.20$	$-\ 72.63\pm.09$	$+0.03\pm.22$	-0.17	5

If we multiply the mean errors given by 0.6745, to reduce them to probable errors, we shall see that only four of the fifteen differences are less than their probable errors. The deviations which call for especial consideration are the following four:

1. The motion of the perihelion of Mercury. The discordance in the secular motion of this element is well known.

2. The motion of the node of Venus. Here the discordance is more than five times its probable error.

3. The perihelion of Mars. Here the discordance is three times its probable error.

4. The eccentricity of Mercury. The discordance is more than twice its probable error. It is to be remarked, however, that the probable error of this quantity is very largely a matter of judgment, and that its value may have been underestimated.

The deviations, if not due to erroneous masses, may be explained on two hypotheses. One is that propounded by Prof. HALL,* that the gravitation of the Sun is not exactly as the inverse square, but that the exponent of the distance is a fraction greater than 2 by a certain minute constant. This hypothesis accounts only for the motions of the perihelia, and not for any other discordances.

The other hypothesis is that of the action of unknown masses or arrangements of matter. Since the latter hypothesis

---

would account for other motions than those of the perihelia, it might seem that the existence of the other discordances tells very strongly in its favor. The hypotheses of possible dis- tributions of unknown matter, therefore, have first to be con- sidered.*

Hypothesis of non-sphericity of the Sun.

56. In a case where our ignorance is complete, all hypotheses which do not violate known facts are admissible. Beginning at the center and passing outward, the first question arises whether the action may not be due to a non-spherical distri- bution of matter within the body of the Sun, resulting in an excess of its polar over its equatorial moment of inertia. The theory of the Sun which has in recent times been most gener- ally accepted is that its interior may be regarded as gaseous, or rather as a form of matter which combines the elasticity and mobility of a gas with the density of a liquid. Such being the case, we may conceive that vortices of which the axes coincide with that of rotation may exist in the interior in such a way that the surfaces of equal density are non- spherical. A very small inequality of this sort would suffice to account for the motion of the perihelion of Mercury.

This hypothesis admits of an easy test. Whatever be the nature or amount of the inequality, a simple computation shows that to account for the observed phenomenon it is necessary and sufficient that the equipotential surfaces at the surface of the Sun should have an ellipticity of rather more than half a second of arc. It can not, I conceive, be doubted that the visible photosphere is an equipotential surface. We have then to inquire whether there is any such ellipticity of the photosphere as that required by the hypothesis. This question seems completely set at rest by the great mass of heliometer measures made by the German observers in con- nection with the transits of Venus of 1874 and 1882, which have been discussed by Dr. AUWERS. The general result is

---

*After carrying out the investigations of this chapter, I find that the subject was studied on similar lines by Dr. P. HARZER nearly three years ago, and that I made certain suggestions on the subject to Dr. BAUSCH- INGER ten years ago. See *Astrononische Nachrichten*, Vol. 109, p. 32, and Vol. 127, p. 81.

that the mean of the equatorial measures are slightly less than the mean of the polar measures, the difference, however, being within the probable errors of the results. I conclude that there can be no such non-symmetrical distribution of matter in the interior of the Sun as would produce the observed effect.

This same conclusion seems to apply to matter immediately around the photosphere. An equatorial ring of planetoids, or gaseous substances of the required mass, very near the photosphere, would render the equipotential surfaces of the photosphere elliptical to a degree which seems precluded by the measures in question. At a very short distance from the surface, however, the effect would be inappreciable.

Hypothesis of an intra-mercurial ring or group of planetoids.

57. Passing outward, we have next to consider the hypothesis of an intra-mercurial ring adequate to produce the observed phenomena. In a first approximation we may suppose the ring circular. Its mass can not be determined, because it will depend upon the distance; we have to determine a certain function of the mass and distance adequate to produce the observed motion of the perihelion. Then we must inquire what effect the ring will have on the secular variations of the other elements, both of Mercury and of the other planets, and see if these effects can be reconciled with observation. In the computations I have assigned to the excess of motion the provisional value $40''.7$. If the ring is not very distant from the Sun the motion which it will produce in the perihelion of a planet whose mean motion is n and whose mean distance is a may be represented in the form

$$D_t \pi = \frac{\mu n}{a^2}$$

μ being a function of the mass of the ring and of its radius, which is nearly the same for all of the planets, so long as the radius of the ring is only a small fraction of the distance of Mercury. A first approximation to μ is—

$$\mu = \frac{3}{4} m r^2$$

m being the ratio of its mass to that of the Sun and r its radius. Multiplying these motions in the case of the four planets by their eccentricities, we find that the hypothetical ring will produce the following secular variations:

$$\text{Mercury,} \quad D_t \pi = 40.7''; \quad e D_t \pi = 8.38''$$

Venus,	4.6	0.031
Earth,	1.5	0.025
Mars,	0.34	0.031

Owing to the smallness of the eccentricities the effect is insensible, except in the case of Mercury, so that the ring will not account for the observed excess of motion of the perihelion of Mars.

Such a ring will necessarily produce a motion of the plane of the orbit of Mercury or Venus, or of both, because it can not lie in the plane of both orbits.

Let us put i_1 for its inclination to the ecliptic, and θ_1 for the longitude of its node on the ecliptic; and let us put, also,

$$p_1 = i_1 \sin \theta_1$$
$$q_1 = i_1 \cos \theta_1$$

and let $p, p', \ldots, q, q', \ldots$ be the corresponding quantities for the planets. The theory of the secular variations then shows that the ring will produce a motion of the plane of the orbit of Mercury given by the equations

$$D_t p_1 = \frac{\mu n}{a^2} (q_1 - q) = 40''.7 \, (q_1 - q)$$

$$D_t q_1 = \frac{\mu n}{a^2} (p - p_1) = 40''.7 \, (p - p_1)$$

Expressing the motions of p and q in terms of the motions of i and θ, which is necessary, owing to the very different weights of the determination of the motion of the planes of Mercury and Venus in the direction of these two coordinates, we have the following expressions for these two motions, which we equate to the observed excesses:[*]

$$
\begin{aligned}
-4.96 + 26.9\,q_1 + 28.4\,p_1 &= +0.57 \pm 0.50 \\
-0.27 + 0.8 + 3.0 &= +0.63 \pm 0.12 \\
0.00 + 28.4 - 26.9 &= +0.50 \pm 0.80 \\
0.00 + 3.0 - 0.8 &= +0.45 \pm 0.30 \\
0.00   - 1.5 &= -0.25 \pm 0.25
\end{aligned}
$$

Multiplying the conditional equations thus formed by such factors as will make the mean error of each equation nearly $\pm\,0''.5$, we have the following conditional equations for p_1 and q_1:

$$
\begin{aligned}
27\,q_1 + 28\,p_1 &= +5.53 \\
3 + 12 &= +3.60 \\
17 - 16 &= +0.30 \\
5 - 1 &= +0.77 \\
0 - 3 &= -0.50
\end{aligned}
$$

The solution of these equations gives very nearly

$$
\begin{aligned}
q_1 &= +0.11; & i_1 &= 9^\circ \\
p_1 &= +0.12; & \theta_1 &= 48
\end{aligned}
$$

This great inclination seems in the highest degree improbable if not mechanically impossible, since there would be a tendency for the planes of the orbits of a ring of planets so situated to scatter themselves around a plane somewhere between that of the orbit of Mercury and that of the invariable plane of the planetary system, which is nearly the same as that of the orbit of Jupiter. Moreover, the motion of the perihelion of Mars is still unaccounted for and that of the node of Venus only partially accounted for, as shown by the large residual of the second equation. In fact, the great inclination assigned to the ring comes from the necessity of representing as far as possible the latter motion.

---

[*] It will be noticed that in forming these equations I have neither used the final values of the absolute terms, nor taken account of the fact that the observed motions of the planes are referred to the ecliptic. Changes thus produced in the equations are too minute to affect the conclusion.

There would of course be no dynamical impossibility in the hypothesis of a single planet having as great an inclination as that required. But I conceive that a planet of the adequate mass could not have remained so long undiscovered. Whether we regard the matter as a planet or a ring, a simple computation shows that its mass, if at the Sun's surface, would be about $\frac{1}{1050}$ that of the Sun itself, and one-fourth of this if at a distance equal to the Sun's radius. We may conceive, if we can not compute, how much light such a mass of matter would reflect. Altogether, it seems to me that the hypothesis is untenable.

Hypothesis of an extended mass of diffused matter like that which reflects the zodiacal light.

58. The phenomenon of the zodiacal light seems to show that our Sun is surrounded by a lens of diffused matter which extends out to, or a little beyond, the orbit of the Earth, the density of which diminishes very rapidly as we recede from the Sun. The question arises whether the total mass of this matter may not be sufficient to cause the observed motion.

So far as the action of that portion of matter which is near the Sun is concerned, the conclusions just reached respecting a ring surrounding the Sun will apply unchanged, because we may regard such a mass as made up of rings. Observation seems to show that the lens in question is not much inclined to the ecliptic, and if so it would produce a motion of the nodes of Venus and Mercury the opposite of that indicated by the observations.

There is another serious difficulty in the way of the hypothesis. A direct motion of the perihelion of a planet may be taken as indicating the fact that the increase of its gravitation toward the Sun as it passes from aphelion to perihelion is slightly greater than that given by the law of the inverse square. This increase would be produced by a ring of matter either wholly without or wholly within the orbit. But if we suppose that the orbit actually lies in the matter composing such a ring, the effect is the opposite; gravitation toward the

Sun is relatively diminished as the planet passes from aphelion to perihelion, and the motion of the perihelion would be retrograde.

It can not be supposed that that part of the zodiacal light more distant from the Sun than the aphelion of Mercury is even as dense as that portion contained between the aphelion and the perihelion distances. The result in question must therefore be due wholly to that part of the matter which lies near to the Sun, and we thus have all the difficulties of the intra-mercurial ring theory, with one more added.

Hypothesis of a ring of planetoids between the orbits of Mercury and Venus.

59. It appears that any ring or zone of matter adequate to produce the observed effect must lie between the orbits of Mercury and Venus. Its assignment to this position requires a more careful determination of its possible eccentricity. There will be six independent elements to be determined; the mass, the mean distance, the eccentricity, the perihelion, the inclination, and the node.

I find that the observed excesses of motion of the elements of Mercury and Venus will be approximately represented by elements not differing much from the following:

Total mass of group	$\dfrac{1}{37\,000\,000}$
Mean distance	0.48
Eccentricity of orbit	0.04
Longitude of perihelion	$10°$
Longitude of node	$35°$
Inclination to ecliptic*	$7°.5$
Probable diameter at distance unity if agglomerated into a single planet .	$3''.5$

Considerations on the admissibility of the hypothesis—Possible mass of the minor planets.

60. Although the preceding hypothesis is that which best represents the observations of Mercury and Venus, we can not, in the present condition of knowledge, regard it as more than a curiosity. True, it is plausible at first sight. Since,

as already remarked, any disturbing body of sufficient mass
to cause the observed excess of motion of the perihelion of
Mercury would change the position of the planes of the orbits,
and since observations give apparent indications of such a
change in the plane of the orbit of Venus, it might appear
that we have here a very good ground for the view that all
the motions are due to the attraction of unknown masses.
But the great difficulty is that the excess of motion of the
orbital planes is in the opposite direction from what we should
expect. A group of bodies revolving near the plane of the
ecliptic would produce a retrograde motion of the nodes. But
the observed excess is direct. A direct motion can be produced only in case the orbits are more inclined than those of
the disturbed planet. In admitting such orbits we encounter
difficulties which, if not absolutely insurmountable, yet tell
against the probability of the hypothesis.

The hypothesis carries with it the probable result that the
excess of motion of the perihelion of Mars is produced by the
action of the minor planets. I have considered the question
of this action in an unpublished investigation. From the probable albedo and magnitude of the minor planets and the observations of BARNARD and others on their diameters, I have
determined the probable mass of each part of the group having
a given opposition magnitude. The result is that the number
of these bodies having such a magnitude appears to progress
in a fairly uniform manner through several magnitudes. The
ratio of progression may lie anywhere between the limits 2
and 3. Up to the limit 3 the total mass, if continued on to
infinity, could not produce any appreciable effect on the motion
of Mars. But if we suppose a larger ratio than 3 to prevail,
then the number of planets of smaller magnitude would be so
numerous as to form a zone of light across the heavens, as may
readily be seen by considering that the total amount of light
reflected from the planets of each order of magnitude would
form an increasing series, since the ratio between the brillian-
cies of two objects of unit difference in magnitude is only
about 2.5. We may therefore suppose that the faint band of
light which is said to be visible across the entire heavens as
a continuation of the zodiacal light, as well as the "gegen-

schein," is due to these minute bodies, and yet find their total mass too small to produce any appreciable effect.

Whether we can assign to the components of such a group any magnitude so small that they would be individually invisible, and a number so small that they would not be seen collectively as a band of light brighter than the zodiacal arch, and yet having a total mass so large as to produce the observed effects, is a very important question which can not be decided without exact photometric investigations. It is, however, certain that if we could do so we should have to suppose a very unlikely discontinuity in the law of progression between each magnitude and the number of bodies having that magnitude. It must therefore suffice for our present object that we regard the hypothesis of such bodies as unsatisfactory.

Hypothesis that gravitation toward the sun is not exactly as the inverse square of the distance.

61. Prof. HALL's hypothesis seems to me provisionally not inadmissible. It is, that in the expression for the gravitation between two bodies of masses m and m' at distance r

$$\text{Force} = \frac{m\,m'}{r^n}$$

the exponent n of r is not exactly 2, but $2 + \delta$, δ being a very small fraction. This hypothesis seems to me much more simple and unobjectionable than those which suppose the force to be a more or less complicated function of the relative velocity of the bodies. On this hypothesis the perihelion of each planet will have a direct motion found by multiplying its mean motion by one-half the excess of the exponent of gravitation.

Putting

$$n = 2.000\ 000\ 1574$$

the excess of motion of each perihelion of the four inner planets would be as follows. It will be seen that the evidence in the case of Venus and the Earth is negative, owing to the

very small eccentricities of their orbits, while the observed
motion in the case of Mars is very closely represented.

	$D_t\pi$	$eD_t\pi$
	''	''
Mercury,	42.34	8.70
Venus,	16.58	0.11
Earth,	10.20	0.17
Mars,	5.42	0.51

An independent test of this hypothesis in the case of other
bodies is very desirable. The only case in which there is any
hope of determining such an excess is that of the Moon, where
the excess would amount to about 140'' per century. This is
very nearly the hundred-thousandth part of the total motion
of the perigee. The theoretical motion has not yet been computed with quite this degree of precision. The only determination which aims at it is that made by HANSEN.* He finds

	Theory.	Obser.	Diff.
	''	''	''
Annual mot. of perigee,	146 434.04;	146 435.60;	+1.56
Annual mot. of node,	−69 676.76	−69 679.62;	−2.86

The observed excess of motion agrees well with the hypothesis, but loses all sustaining force from the disagreement in
the case of the node. The differences HANSEN attributes
(wrongly, I think) to the deviation of the figure of the Moon
from mechanical sphericity.

*Consistency of Hall's hypothesis with the general results of the
law of gravitation.*

62. The law of the inverse square is proven to a high degree
of approximation through a wide range of distances. The close
agreement between the observed parallax of the Moon and
that derived from the force of gravitation on the Earth's surface shows that between two distances, one the radius of the
Earth and the other the distance of the Moon, the deviation
from the law of the square can be only a small fraction of the

---

* *Darlegung, etc.: Abhandlungen der Math.-Phys. Classe der Kön. Sächsi-
schen Gesellschaft der Wissenschaften*, VI, p. 348.

thousandth part, or, we may say, a quantity of the order of magnitude of the five-thousandth part.

Coming down to smaller distances, we find that the close agreement between the density of the Earth as derived from the attraction of small masses, at distances of a fraction of a meter, with the density which we might *a priori* suppose the Earth to have, shows that within a range of distance extending from less than one meter to more than six million meters, the accumulated deviation from the law can scarcely amount to its third part. The coincidence of the disturbing force of the Sun upon the Moon with that computed upon the theory of gravitation, extends the coincidence from the distance of the Moon to that of the Sun, while KEPLER'S third law extends it to the outer planets of the system. Here, however, the result of observations so far made is relatively less precise. We may therefore say, with entire confidence, as a result of accurate measurement, that the law of the inverse square holds true within its five-thousandth part from a distance equal to the Earth's radius to the distance of the Sun, a range of twenty-four thousand times; that it holds true within a third of its whole amount through the range of six million times from one meter to the Earth's radius; and within a small but not yet well-defined quantity from the distance of the Sun to that of Uranus, in which the multiplication is twentyfold.

If HALL'S hypothesis contradicted these conclusions it would be untenable. But a very simple computation will show that, assuming the force to vary as $r^{-(2+\delta)}$, δ being a minute constant sufficient to account for the motion of the perihelion of Mercury, the effect would be entirely inappreciable in the ratio of the gravitation of any two bodies at the widest range of distance to which observation has yet extended. Although the total action of a material point on a spherical surface surrounding it would converge to zero when the radius became infinite, instead of remaining constant, as in the case of the inverse square, yet the diminution in the action upon a surface no larger than would suffice to include the visible universe would be very small.

*Masses of the planets which represent the secular variations of
other elements than the perihelia.*

63. On HALL's hypothesis the secular variations of all the
elements other than the perihelia will remain unchanged.

Our next problem is to consider the possibility of representing the variations of the other elements by admissible masses
of the known planets. In § 55 I have given a comparison of
the secular variations as they result from observations, with
their theoretical expressions in terms of corrections to a certain system of masses. When the equations thus formed are
multiplied by the factors $\sqrt{w}$, which make the mean error of
each equation unity, we have the following system of equations, in which we put $\nu = 10x$:

$0x$	$+ 6\nu'$	$+ 2\nu''$	$+ 0\nu'''$	$= -1.7$	$r = -1.8$
0	$- 1$	$- 2$	0	$= +0.5$	$+0.5$
$- 7$	-108	$- 27$	$- 3$	$= +0.5$	$+1.1$
-65	0	$- 24$	$- 1$	$= +0.6$	$+0.7$
$+25$	0	0	$- 1$	$= +1.5$	$+1.3$
$+21$	-234	-346	-10	$= +4.2$	0.0
-12	$+ 13$	0	-16	$= +0.2$	$+0.1$
$- 9$	-123	0	$- 3$	$= -2.0$	-0.7
$+ 1$	0	$+ 8$	0	$= +1.1$	$+1.3$
$- 2$	$+ 60$	0	0	$= +0.4$	-0.2
-14	-126	$- 37$	$- 5$	$= -0.8$	-0.2

The resulting normal equations are

$$5766\,x - 1563\,\nu' - 4991\,\nu'' + 140\,\nu''' = + 114$$
$$- 1563 + 101231 + 88556 + 3455 = - 670$$
$$- 4991 + 88556 + 122462 + 3750 = - 1446$$
$$+ 140 + 3455 + 3750 + 401 = - 39$$

Along with the results of the solution of these equations I
place, for comparison, the values of Chapter V, which have
been considered most probable.

	From sec. var.		From other sources.	
$10x = \nu$	$= + 0.070$		$+ 0.08$	± 0.20
ν'	$= + 0.0100$	$\pm .0056$	$+ 0.0084$	± 0.0028
ν''	$= - 0.0183$	$\pm .0052$	$- 0.00304$	± 0.0015
ν'''	$= - 0.0115$	$\pm .067$	$+ .0037$	± 0.018

By substitution in the conditional equations we find for the mean error corresponding to weight unity—

$$\epsilon_1 = \pm 1.14$$

In forming these equations they were reduced by multiplication to a supposed mean error of ± 1. Speaking in a general way we may therefore say that the representation of the secular variations, those of the perihelia being ignored, by these corrections to the masses is satisfactory. Except for the large discordance in the motion of the eccentricity of Mercury the mean error would have been less than unity.

Comparing the two sets of values we find that the masses of Mercury, Venus, and Mars agree well with those derived from other sources. Very different is it with the mass of the Earth. The discordance is here more than the hundredth part of its whole amount, which involves a discordance of more than the three-hundredth part in the value of the solar parallax. Let us now proceed in the reverse order, and determine the value of the solar parallax from the mass of the Earth, as derived from the preceding data.

Preliminary adjustment of the two sets of masses.

64. We make the best adjustment for this purpose by adding to the equations of condition last given the additional ones derived from the values of the masses discussed in Chapter V. Multiplying each value of ν by the factor necessary to reduce the mean error of the second member of the equation to unity, we have the following conditional equations:

$$50\, x = + 0.4$$
$$300\, \nu' = + 2.9$$
$$50\, \nu''' = \quad 0.0$$
$$30\, \nu''' = + 0.42$$

Of the last two equations it may be remarked that the first is that given by Prof. HALL's original mass of 1877, while the last is derived by Dr. HARSHMAN from HALL's observations of the outer satellite made during the opposition of 1892.

When we add to the normal equations already formed the products of these last equations by the factors of the unknown quantities, the system of normal equations is as follows:

$$
\begin{array}{rrrrr}
8266\,x & -\ 1563\,\nu' & -\ 4991\,\nu'' & +\ 140\,\nu''' & = +134 \\
-1563 & +230831 & +\ 88556 & +3455 & = +374 \\
-4991 & +\ 88556 & +122462 & +3750 & = -1446 \\
+\ 140 & +\ 3455 & +\ 3750 & +3801\ \cdot & = -26
\end{array}
$$

The solution of these equations gives the following values of the unknown quantities:

$$
\begin{aligned}
x &= +\,0.0071 \pm .0120 \\
\nu &= +\,0.071\ \ \pm .120 \\
\nu' &= +\,0.0084 \pm .0024 \\
\nu'' &= -\,0.0177 \pm .0035 \\
\nu''' &= +\,0.0027 \pm .016
\end{aligned}
$$

Here again we note that, the Earth aside, the results for the masses are quite satisfactory. The correction to Prof. HALL'S original mass of Mars is so minute and so much less than its probable error that we may consider this value of the mass to be confirmed, and may adopt it as definitive without question. The corrections to the masses of Mercury and Venus are scarcely changed. The mean residual is reduced to

$$
\epsilon = \pm\,0.91
$$

which is less than the supposed value.

We have, therefore, so far as these results go, no reason for distrusting the following value of the solar parallax, which results from that of the mass of the Earth thus derived:

$$
\pi = 8''.759 \pm ''.010
$$

The critical examination and comparison of this and other values of the parallax is the work of the next two chapters.

CHAPTER VII

VALUES OF THE PRINCIPAL CONSTANTS WHICH DEFINE THE MOTIONS OF THE EARTH.

The Precessional Constant.

65. The accurate determination of the annual or centennial motion of precession is somewhat difficult, owing to its dependence on several distinct elements, and to the probable systematic errors of the older observations in Right Ascension and Declination. What is wanted is the annual motion of the equinox, arising from the combined motions of the equator and the ecliptic, relative to directions absolutely fixed in space. As observations can not be referred to any line or plane which we know to be absolutely fixed, we are obliged to assume that the general mean direction of the fixed stars remains unchanged, or, in other words, that the stellar system in general has no motion of rotation. This is a safe assumption so far as the great mass of stars of smaller magnitude is concerned. But it is not on such stars that we have the earliest accurate observations. Moreover, observed Right Ascensions of these fainter stars relative to the brighter ones are subject to possible systematic errors, arising from the personal equation being different for brighter and fainter stars. In the case of the stars observed by BRADLEY, there is frequently such community of proper motion among neighboring stars that we can not be quite sure that all rotation is eliminated in the general mean. Under these circumstances we have only to make the best use that we can of existing material.

We must also remember that observed Right Ascensions are not directly referred to the equinox, but to the Sun, of which the error of absolute mean Right Ascension must be determined. This again can be done only from observed declinations, since by definition the equinox is the point at which the Sun crosses the equator. It is also to be noted that the clock stars which are directly compared with the Sun by no

124

means include the whole list to be used as absolute points of reference. We therefore have three separate steps in determining completely a correction to the adopted annual precession:

(1) The correction to the Sun's absolute mean Right Ascension or longitude.

(2) The correction to the general mean Right Ascension of the clock stars relative to the Sun.

(3) The determination of the clock stars relative to the great mass of stars.

It goes without saying that the determinations of these three quantities are entirely independent of each other, and that the precision of the result depends on the precision of each separate determination.

The motion of the pole of the equator, on which the lunisolar precession depends, may be determined by observed Declinations quite independently of the Right Ascensions. A determination of the precession from the latter includes the planetary precession, but as this has to be determined from theory independently of observations, we have, in observed Right Ascensions and Declinations, two independent methods of determining the motion of the equator.

It fortunately happens that the constant of precession is not so closely connected with other constants that a small error in its determination will seriously affect our general conclusions, or the reduction of places of the fixed stars, because the effect of an error will be nearly eliminated through the proper motions of the fixed stars, or the motions of the planets in longitude. I have therefore satisfied myself with reviewing and combining the four best determinations.

I pass over in silence the classic determinations of BESSEL and OTTO STRUVE, because the material on which they depend has been incorporated in more recent works. Of these the one which seems entitled to most weight is that of LUDWIG STRUVE, *Bestimmung der Constante der Præcession, und der eigenen Bewegung des Sonnensystems.** This work was suggested by the completion of AUWERS' re-reduction of BRADLEY'S Observations, and of the Pulkowa standard catalogues for 1845,

---

* Mémoires de l'Académie Impériale des Sciences de St. Pétersbourg. VIIe Série. Tome xxxv, No. 3.

1855, and 1865. It depends entirely on the BRADLEY stars, and the result, when reduced to the most probable equinox, may be regarded as the best now derivable from those stars, or, at least, as not susceptible of any large correction.

He, of course, includes in his work the determination of the motion of the solar system relative to the mass of the stars. In addition to this, the possibility of a common rotation of the BRADLEY stars around the axis of the Milky Way is considered. This rotation I should be disposed to regard as zero for the present.

In place of considering each of the 2,509 stars singly, he divides the celestial sphere into 120 spherical trapezoids, each covering 15 degrees in Declination, and an arc of Right Ascension equal approximately to one hour of a great circle at the equator. The question might be legitimately raised whether a different system of weighting the trapezoids, founded on a consideration and comparison of the proper motions in Right Ascension and Declination would not have been advisable. I am, however, fairly confident that no change in this respect would have materially affected the result. With this work of STRUVE I have combined those of BOLTE, DREYER, and NYRÉN.

In the case of the Right Ascensions it is necessary to reduce all the results to the equinox determined in the last chapter. From this chapter it appears that the standard Right Ascensions with which the reduction of the preceding investigations have been made require a correction to the centennial motion of $+ 0''.30$. Reducing each determination to the equinox thus defined, we have the following results for the general precession in Right Ascension at the epoch 1800:

L. STRUVE, from the comparison of AUWERS-BRADLEY with the modern Pulkowa Right Ascensions . . . $m = 46''.0301$; $w = 4$

DREYER, from the comparison of LALANDE's Right Ascensions with those of SCHELLERUP . . . . . $46\ .0611$; $w = 2$

NYRÉN, by the comparison of BESSEL's Right Ascensions with those of SCHJELLERUP . . . . $46\ .0456$; $w = 1$

Mean . . . . $46\ .0526$

The weights here assigned are of course a matter of judgment. The general agreement of the results is as good as we could expect.

From observed declinations we have—

L. STRUVE, from the comparison of
AUWERS - BRADLEY with modern
Pulkowa catalogues . . . . . . $n = 20''.0495;\ w = 2$
BOLTE, from the comparison of LA-
LANDE'S Declinations with those of
SCHJELLERUP . . . .   20 .0537; $w = 1$
Mean . . . .   20 .0509

We have now to combine these independent results. I propose to call *Precessional Constant* that function of the masses of the Sun, Earth, and Moon, and of the elements of the orbits of the Earth and Moon, which, being multiplied by half the sine of twice the obliquity, will give the annual or centennial motion of the pole on a great circle, and being multiplied by the cosine of the obliquity will give the luuisolar precession at any time. It is true that this quantity is not absolutely constant, since it will change in the course of time, through the diminution of the Earth's eccentricity. This change is, however, so slight that it can become appreciable only after several centuries. If, then, we put

P, the precessional constant, we have, for the annual general precession in Right Ascension and Declination—

$$m = \text{P} \cos^2 \varepsilon - \varkappa \sin \text{L} \cosec \varepsilon$$
$$n = \text{P} \sin \varepsilon \cos \varepsilon$$

L being the longitude of the instantaneous axis of rotation of the ecliptic, and $\varkappa$ its annual or centennial motion. From the definitive obliquity and masses of the planets adopted hereafter, we find the following values of $\varkappa$, L, and ε, for 1800 and 1850:

	1800.	1850.
$\log \varkappa =$	1.67372;	1.67341
$\text{L} =$	173° 2'.31;	173° 29'.68
$\varepsilon =$	23 27 .92;	23 27 .53

We thus find the following values of P, the unit of time being 100 solar years:

From Right Ascensions,	$P = 5490.12$; $w = 2$
From Declinations,	$P = 5489.44$; $w = 1$

Mean, $P = 5489''.89$

As the data used in STRUVE'S investigation may be considered of a more certain kind than those used by the others, we may compare these results with those which follow from STRUVE'S work alone. They are

From Right Ascensions,	$P = 5489.83$
From Declinations,	$P = 5489.06$

Giving double weight to the results from the Right Ascensions, the results may be expressed as follows:

From STRUVE'S investigation,	$P = 5489.57$
From the other two works,	$P = 5490.18$

Before concluding this investigation, I had adopted as a preliminary value

$$P = 5489''.78$$

As this result does not differ from the one I consider most probable, $5489''.89$, by more than the probable error of the latter, and diverges from it in the direction of the best determination, I have decided to adhere to it as the definitive value.

The centennial value of P is subjected to a secular diminution of $0''.00364$ per century, owing to the secular diminution of the eccentricity of the Earth's orbit. We therefore adopt

$$P = 5489.78 - 0.00364 \, T \text{ for a tropical century.}$$
$$P = 5489.90 - 0.00364 \, T \text{ for a Julian century.}$$

In the use of P I at first neglected the secular variation, but have added its effect to the results developed in powers of the time.

Constant of nutation derived from observations.

66. The determination of this constant from observations is extremely satisfactory, owing to the completeness with which systematic errors may be eliminated. If, with a meridian instrument, regular observations are made through a draconitic period, on a uniform plan, upon stars equally distributed through the circle of Right Ascension, the observations being made daily through more than 12 hours of Right Ascension, all systematic errors in the determination of the nadir point and all having a diurnal or annual period may be completely eliminated from the constant in question. These conditions are so nearly fulfilled in the observations with the Greenwich transit circle, and, to a less extent, in those with the Washington transit circle, that the results of the work with those two instruments alone are entitled to greater weight than has hitherto been supposed. I have, however, discussed quite fully all previous determinations of which it seemed that the probable mean error would be less than $\pm\ 0''.10$.

Referring to the volume on the subject to be hereafter published, the results of the discussion are presented in the following table. The weights are assigned on the supposition that weight unity should correspond to a mean error of about $\pm\ 0''.07$, or to a probable error of $\pm\ 0''.05$, this probable value being not entirely a matter of computation from the discordance of the separate results, but, to a certain extent, a matter of judgment.

It must be understood that the results below are not always those given by the authors who are quoted, but that their discussion has, wherever possible, been subjected to a revision by the introduction of modern data, or by what seemed to me improved combinations. Thus, NYRÉN's equations have been reconstructed on a system slightly different from his, and have been corrected for CHANDLER's variation of latitude. PETERS's classical work has also been corrected by the introduction of later data, and by a re-solution of his equations. The Greenwich and Washington results have been derived from the discussion in *Astronomical Papers*, Vol. II, Part VI.

Values of the constant of nutation derived from observations.

	N	$wt.$
BUSCH, from BRADLEY's observations with the zenith sector	9.232	1
ROBINSON, from Greenwich mural circles	9.22	1
PETERS, from Right Ascensions of Polaris	9.214	4
LUNDAHL, from Declinations of Polaris	9.236	1.5
NYRÉN, from v Urs. Maj.	9.254	3
" " o Draconis	9.242	2.5
" " ι Draconis	9.240	4
DE BALL, from WAGNER's Right Ascensions of Polaris	9.162	3
DE BALL, from WAGNER's Declinations of Polaris	9.213	3
DE BALL, from WAGNER's Right Ascensions of 51 Cephei	9.252	3
DE BALL, from WAGNER's Declinations of 51 Cephei	9.227	3
DE BALL, from WAGNER's Right Ascensions of δ Urs. Min.	9.208	3
DE BALL, from WAGNER's Declinations of δ Urs. Min.	9.263	3
Greenwich North-Polar Distances of Southern Stars, Series I	9.116	3
Greenwich North-Polar Distances of Southern Stars, Series II	9.201	3
Greenwich North-Polar Distances of Northern Stars, Series I	9.204	4
Greenwich North-Polar Distances of Northern Stars, Series II	9.223	4
Washington Transit Circle, southern stars	9.217	6
" " " northern stars	9.177	3
Greenwich, Right Ascensions of Polaris	9.153	2
" Declinations of Polaris	9.242	2
" Right Ascensions of 51 Cephei	9.135	2
" Declinations of 51 Cephei	9.162	2
" Right Ascensions of δ Urs. Min.	9.147	2
" Declinations of δ Urs. Min.	9.235	2
" Right Ascensions of λ Urs. Min.	9.161	1
" Declinations of λ Urs. Min.	9.339	1
Mean	9.210;	$wt. = 72$

The mean error corresponding to weight unity when derived from the discordance of the results is ± 0″.008, while the estimate was ± 0″.070. We may therefore put, as the result of observation—

$$N = 9''.210 \pm 0''.008$$

Relations between the constants of precession and nutation, and the quantities on which they depend.

67. The formulæ of precession and nutation have been developed by OPPOLZER with very great rigor and with great numerical completeness as regards the elements of the Moon's orbit, in the first volume of his *Bahnbestimmung der Kometen und Planeten*, second edition, Leipzig, 1882. What is remarkable about this work is that it constantly takes account of the possible difference between the Earth's axis of rotation and its axis of figure, a distinction which has become emphasized by CHANDLER'S discovery since OPPOLZER wrote. His theory however fails to take account of the change in the period of the Eulerian nutation produced by the mobility of the ocean and the elasticity of the Earth. But this effect is of no importance in the present discussion.

From OPPOLZER'S developments, I have derived the following expressions, in which the numerical coefficients may be regarded as absolute constants, so accurately determined that no question of their errors need now be considered. These results have been derived quite independently of the similar ones by Mr. HILL in the *Astronomical Journal*, Vol. XI, which are themselves independent of OPPOLZER'S work. In these formulæ we have—

N, the constant of lunar nutation of the obliquity of the ecliptic, as defined by the equation $\varDelta\varepsilon = N \cos \Omega$, and expressed in seconds of arc;

P, so much of the precession of the equinox on the fixed ecliptic of the date, in seconds of arc and in a Julian year, as is due to the action of the Moon;

P′, so much of the same precession as is due to the action of the Sun.

We thus have,

$$\text{luni-solar precession} = P + P'$$

ε, the obliquity of the ecliptic;

μ, the ratio of the mass of the Moon to that of the Earth;

A, the mean moment of inertia of the Earth relative to axes passing through its equator;

C, the same moment relative to its polar axis.

With these definitions we have,

General value.   Special value for 1850.

$$N = [5.40289] \; \cos \varepsilon \; \frac{\mu}{1+\mu} \frac{C-A}{C} = [5.36542] \; \frac{\mu}{1+\mu} \frac{C-A}{C}$$

$$P = [5.975052] \cos \varepsilon \; \frac{\mu}{1+\mu} \frac{C-A}{C} = [5.937585] \frac{\mu}{1+\mu} \frac{C-A}{C}$$

$$P' = [3.72509] \; \cos \varepsilon \; \frac{C-A}{C} = [3.68762] \; \frac{C-A}{C}$$

The special values for 1850 are found by putting for the value of the obliquity of the ecliptic for 1850,

$$\varepsilon = 22^\circ \; 27' \; 31''.7$$

The mass of the Moon from the observed constant of nutation.

68. From the two quantities given by observation, N and $P + P' = p_0$, these equations enable us to determine the two unknown quantities μ and $\frac{C-A}{C}$. As the easiest way of showing the uncertainty of the Moon's mass, arising from uncertainty of the precession and nutation, I give the value of its reciprocal corresponding to different values of these quantities in the following table:

Reciprocals of the mass of the Moon corresponding to different values of the nutation-constant and luni-solar precession.

p_0	$N = 9''.20$	$N = 9''.21$	$N = 9''.22$
''			
50. 35	81. 81	81. 53	81. 25
50. 36	81. 86	81. 58	81. 30
50. 37	81. 91	81. 63	81. 35

Taking for the constant of nutation the value just found,

$$N = 9''.210 \pm ''.068$$

and for the luni-solar precession,

$$p_0 = 50''.36 \pm ''.006$$

we have, for the reciprocal of the mass of the Moon and its mean error:

$$\frac{1}{\mu} = 81.58 \pm 0.20$$

The Constant of Aberration.

69. In the determination of astronomical constants the investigation of the constant of aberration necessarily takes a very
important place, not only on its own account but on account of
its intimate connection with the solar parallax. A general
determination, founded on all the data available, was therefore
commenced by me as far back as 1890, before the fact of the
variation of terrestrial latitudes had been well established.
The successive discoveries of the law of this variation by
CHANDLER required such alterations in the work as it went
along that much of it is now of too little value for publication
in full. Happily the necessity for a new discussion of the best
determinations at Pulkowa has been done away with by the
papers of CHANDLER himself in the *Astronomical Journal*.

Quite apart from the disturbing influence of the revolution
of the terrestrial pole upon the determination of the constant
of aberration, this constant is itself the one of which the determination is most likely to be affected by systematic errors.
In this respect it is at the opposite extreme from the constant
of nutation. From the very nature of the case it requires a
comparison of observations at opposite seasons of the year,
when climatic conditions are different. In most cases the
determination must even be made at different times of day.
The effect of aberration on a star, for example, is generally at
one extreme when the star culminates in the morning, and at
the other extreme when it culminates in the evening. The
culminations at opposite seasons of the year are necessarily

associated with culminations at opposite times of the day. Moreover, in observations to determine the constant of aberration from Declination, the stars which give the largest coefficients are, for the northern hemisphere, those near 18^h of Right Ascension. Any error peculiar to the times or seasons at which these stars are observed will therefore affect the result systematically.

Right Ascensions of close polar stars also lead to a value of this constant. But the same difficulty still exists. In this case the maxima and minima of aberration occur when the star culminates at noon and midnight. Not only is the aspect of the star different at the two culminations, but the effect of any diurnal change in the instrument will be transferred to the final result for the aberration.

The prismatic method of LOEWY is free from some of these objections. But its application is extremely laborious, and we have, up to the present time, only two determinations by it, one by LOEWY himself, which is only regarded as preliminary, and one by COMSTOCK, in which a large uncertain correction for personal equation was applied.

Under these circumstances the seeking of results derived by methods of the greatest possible diversity is yet more strongly recommended than in the case of the other astronomical constants. I have therefore used not only the PULKOWA determinations, but all those made elsewhere which it seemed worth while to consider. Notwithstanding the great amount of material added to NYRÉN's paper of 1883, it will be seen that the probable error of the final result at which I have arrived is greater than that which he assigns to his result. This is a natural consequence of combining so many separate determinations. The advantage is, however, that the assigned probable error is more likely to be the real one. It is not to be supposed that any of the systematic errors already indicated would pertain to all observers and to all instruments. The final outcome should be a result in which the discordances of the separate determinations show the probable values of all the actual errors, both accidental and systematic.

Determinations founded on the Right Ascensions of circumpolar stars are not affected by the motion of the terrestrial axis, nor are those founded on declinations of these stars, if only the declinations are observed equally at both culminations. But determinations founded on declinations of stars from upper culmination only are necessarily affected by this cause. If however the stars on which the determination is based extend through the whole circle of Right Ascension the effect of the cause in question may be wholly eliminated by a suitable treatment of the equations of condition. To practically eliminate the injurious effect it is not even necessary to determine the exact law of variation. In fact, if the stars observed are equally scattered in Right Ascension, the effect of the variation will be partially eliminated without taking account of it.

CHANDLER has shown that there are two periodic terms in the variation of latitude, one having a period of one year, the other of four hundred and twenty-seven days. I may remark that this combination is in accord with my theory developed in the *Monthly Notices of the Royal Astronomical Society* for March, 1892. It was there shown that any minute annual change of the position of the principal axis of inertia of the Earth—a change which might be produced by the motion of water, ice, and air on its surface—would appear as an annual term in the latitude, six times as great as its actual amount.

Values of the constant of aberration derived from observations.

70. What I have done since this discovery by CHANDLER has been to reexamine the determinations of the constant of aberration made from time to time, to make such corrections in their bases as seemed necessary, and more especially to determine the correction to be applied to each separate result on account of the periodic term in the latitude. No attempt was made to rework completely the original material, except in the case of the results of the Pulkowa and Washington observations with the prime vertical transit. In the case of the former, however, the preliminary results reached from time to time were so accordant with those of CHANDLER that it is a matter of indifference whether we regard them as belonging to his work or to my own.

Owing to the very different estimates placed by the astronomical world upon the Pulkowa determinations and •those made elsewhere, I have used the former quite apart from the others. The complete discussion of each separate value is too voluminous for the present publication, and is therefore reserved for a more extended future publication. At present it appears sufficient to judge the final result by the general discordance of the material on which it rests, rather than by a separate criticism of each particular case.

In the exhibit of results which follows it is to be remarked that NYRÉN's prime vertical observations do not receive a weight as great, relative to the other Pulkowa determinations, as would be given by their assigned probable errors. The reason of this course is that one can not be entirely confident that the results of any one observer with this instrument are free from constant error arising from differences of personal equation in observing a bright and a faint star. Many of the Pulkowa observations are necessarily made in the morning or evening twilight. In the case of an evening observation the star will therefore be much fainter on account of daylight when it transits over the east vertical than it will when it transits over the west vertical one or two hours later. In the case of morning observations the reverse will be true. It is easy to see that if, in consequence of this difference of aspect, the observer notes the passage of the faint image too late, the effect will be to make the constant of aberration too large. The existence of this form of personal equation, when transits are recorded on the chronograph, is so well known that, had NYRÉN's observations been made in this way, I should not have hesitated to ascribe the large values of his aberration constant to this cause. Although it has never been shown that any such personal equation exists when observations are made by eye and ear, as NYRÉN's were, yet when we consider that we are dealing with quantities amounting only to one or two hundredths of a second of arc, and that a personal equation of this kind, undiscoverable by ordinary investigation, might affect the result by this minute amount, we can not but have at least a suspicion that his values may be slightly too large from this cause.

Separate results for the constant of aberration.

A. Standard Pulkowa determinations:

	Ab.	wt.
	"	
Observations with Vertical Circle; Polaris, by PETERS	20.51	2
Observations with Vertical Circle; 7 miscellaneous stars, by PETERS	20.47	2
Observations with Vertical Circle; 1863–1870, Polaris, by GYLDEN	20.41	2
Observations with Vertical Circle; 1871–1875, Polaris, by NYREN	20.51	2
Observations with Prime Vertical; 1842–1844, by STRUVE	20.48	4
Observations with Prime Vertical; 1879–1880 by NYRÉN	20.52	6
Observations with Prime Vertical; 1875–1879, by NYRÉN	20.53	1
Observations with Vertical Circle; 1863–1873, by GYLDEN and NYRÉN	20.52	2
WAGNER: Transits of three polar stars	20.48	5
From Right Ascensions of Polaris; 1842–1844, by LINDHAGEN and SCHWEIZER	20.50	2

Mean result: 20''.493 ± 0''.011

This result may be regarded as identical with that found by NYRÉN in 1882.

B. Other determinations:

	Ab.	ε	wt.
	"		
AUWERS, from observations with the zenith sector at Kew	20.53	±.12	0.5
AUWERS, from WANSTED observations	20.46	±.12	0.5
PETERS, from BRADLEY'S observations of γ Draconis at Greenwich with zenith sector, 1750–1754	20.67		0.5
BESSEL, from Right Ascensions observed by BRADLEY at Greenwich	20.71	±.071	0.5
LINDENAU, from Right Ascensions of Polaris observed at various observatories between 1750 and 1816	20.45	±.05	3

Separate results for the constant of aberration—Continued.

B. Other determinations—Continued.

	Ab.	ϵ	*wt.*
BRINKLEY, from observations of thirteen stars at Trinity College, Dublin, with the 8-foot circle	20.46	±.10	1
PETERS, from STRUVE'S Dorpat observations of six pairs of circumpolar stars	20.36	±.07	2
RICHARDSON, from observations with the Greenwich mural circles	20.50	±.06	3
PETERS, from Right Ascensions of Polaris at Dorpat	20.41		6
LUNDAHL, from Declinations of Polaris at Dorpat	20.55		5
HENDERSON and McLEAR, from α^1 and α^2 Centauri	20.52	±.10	1
MAIN, from observations with the Greenwich zenith tube	20.20	±.10	1
DOWNING, from observations of λ Draconis with reflex zenith tube	20.52	±.05	4
NEWCOMB, from observations of α Lyræ with the Washington prime vertical transit, 1862–1867	20.46	±0.4	6
NEWCOMB, from Right Ascensions of Polaris observed with the Washington transit circle, 1866–1867	20.55	±.05	3
KÜSTNER, from observations of pairs of stars by the TALCOTT method	20.46		4
PRESTON, from observations with the TALCOTT method at Honolulu, 1891–1892	20.43	±.05	4
LOEWY, from his prismatic method	20.45	±.04	5
COMSTOCK, using LOEWY'S method, slightly modified	20.44		3
KÜSTNER, from MARCUSE'S observations, 1889–1890	20.49	±.018	4
WANACH, from Pulkowa prime vertical observations	20.40	±.015	4

Separate results for the constant of aberration—Continued.

B. Other determinations—Continued.

	Ab.	*wt.*
From Greenwich Right Ascensions of polar stars with the transit circle `.	*"* 20.39	3
BECKER, from observations at Strasburg by the TALCOTT method, 1890–1893	20.47	6
DAVIDSON, from similar observations at San Francisco, 1892–1894 ·	20.48	6

Mean result of B: *Ab. const.* $= 20''.463 \pm 0''.013$

The two results, A and B, differ by $0''.030$, a quantity so much greater than their mean errors as to leave room for a suspicion of constant error in one or both means.

The Lunar Inequality in the Earth's motion.

71. The source of this inequality is the revolution of the center of the Earth around the center of mass of the Earth and Moon. The former center describes an orbit which is similar to that of the Moon around the Earth. Since this orbit is not a Keplerian eclipse, but is affected by all the perturbations of the Moon by the Sun, no such element as a semimajor axis can be assigned to it. Instead of this I take as the principal element of the orbit the coefficient of the sine of the Moon's mean elongation from the sun in the expression for the Sun's true longitude. This element is a function of the solar parallax and of the mass of the Moon, which may be derived from the following expression. Let us put

μ; the ratio of the mass of the Moon to that of the Earth;

r, λ, β; the radius vector, true longitude and latitude of the Moon;

r', λ', β'; the same coordinates of the Sun;

s; the linear distance of the Earth's center from the center of mass of the Earth and Moon.

We then have, for the perturbations of the Sun's geocentric place due to the cause in question:

$$\varDelta \log r' = \frac{s}{r'} \cos \beta \cos (\lambda - \lambda')$$

$$\varDelta \lambda' = \frac{s}{r'} \cos \beta \sin (\lambda - \lambda')$$

$$\varDelta \beta' = \frac{s}{r'} \sin \beta$$

and

$$\frac{s}{r'} = \frac{\mu}{1 + \mu} \frac{r}{r'}$$

I have developed these expressions, putting

$$\pi_0 = 8''.848$$

$$\mu = \frac{1}{81}$$

and taking for the Moon's coordinates the values found by DELAUNAY. Putting

D; the mean value of $\lambda - \lambda'$
g, g'; the mean anomalies of the Moon and Sun, respectively,
u'; the Sun's mean elongation from the Moon's ascending node;

the result for $\varDelta \lambda'$ is

$$
\begin{aligned}
\varDelta \lambda' = \quad & 0.533 \sin D \\
+ \ & 0.013 \sin 3\,D \\
+ \ & 0.179 \sin (D + g) \\
- \ & 0.429 \sin (D - g) \\
+ \ & 0.174 \sin (D - g') \\
- \ & 0.064 \sin (D + g') \\
+ \ & 0.039 \sin (3\,D - g) \\
- \ & 0.014 \sin (D - g - g') \\
- \ & 0.013 \sin 2\,u'
\end{aligned}
$$

This value of the lunar inequality is substantially identical with that computed from the tables and formulæ of LEVER-

RIER'S solar tables. The development of the numbers there given lead to the value $6''.534$ of the principal coefficient.

We have now to find what value of the coefficient is given by observations. The observations I make use of are (1) all the observations of the Sun's Right Ascension from early in the century till 1864; (2) The heliometer observations of Victoria made in 1889 on GILL'S plan and worked up by him.

I had intended to use all the observations of the Sun up to the present time. I found however that those made after 1864 gave, by comparison with the published ephemerides, inadmissible positive corrections to the coefficient. This circumstance gives rise to a strong suspicion that in the process of interpolating the Right Ascensions of the Sun during at least some years after 1864, the inequality in question was rounded off to the amount of several hundredths of a second. The results were therefore entirely omitted.

The results for previous years, when the inequality was computed separately for every day of observation, are:

		ΔP	wt.
		$''$	
Greenwich,	1820–'64;	$-.068$	3.0
Paris,	1801–'64;	$-.050$	0.8
Königsburg,	1820–'45;	$-.054$	1.2
Cambridge,	1828–'58;	$-.047$	2.0
Dorpat,	1823–'38;	$+.160$	0.3
Pulkowa,	1842–'64;	$-.058$	0.5
Washington,	1846–'64;	$.000$	0.2

Mean, $\Delta P = -0''.048 \pm 0''.018$

GILL'S result is given in the *Monthly Notices, Royal Astronomical Society*, for April, 1894 (Vol. LIV, page 350.) It is derived in the following way. In the solar ephemeris which he used for comparison the lunar inequalities were computed rigorously from the coordinates of the Moon, putting

$$\pi = 8''.880$$
$$\mu = 1 \div 83$$

To the coefficient P thus arising he found a correction,

$$\Delta P = + 0''.046$$

The above values of π and μ give, on the theory just developed,

$$P = 6''.400$$

Thus GILL's result is, in effect,

$$P = 6''.446$$

while mine, from observations of the Sun, is

$$6''.533 - 0''.048 = 6''.485$$

I consider that these results are entitled to equal weight, and that we may take, as the result of observation,

$$P = 6''.465 \pm 0''.015$$

Solar parallax from the lunar inequality.

72. With the mass of the Moon already found from the observed constant of nutation,

$$\mu = 1 : 81.58 \ (1 \pm .0025)$$

we may now derive a value of the solar parallax quite independent of all other values. The relation between P, π, and the mass of the Moon is of the general form

$$\mu' P = k \pi$$

where k is a numerical constant, and, for brevity,

$$\mu' = \frac{1 + \mu}{\mu} = \frac{1}{\mu} + 1$$

We have found that the following values correspond to one theory:

$$\pi = 8''.848; \qquad \mu' = 82; \qquad P = 6''.533$$

Hence follows

$$\log k = 1.78207$$

so that we have

$$\mu' P = [1.78207] \, \pi$$

The numerical values P = $6''.465$ and $\mu' = 82.58$ now give

$$\pi = 8''.818 \pm 0''.030$$

Values of the solar parallax derived from measurements of Venus on the face of the Sun during the transits of 1874 ·and 1882, with the heliometer and photoheliograph.

73. I put these determinations into one class because they rest essentially on the same principle. Both consist, in effect, in measures of the distance between the center of Venus and the center of the Sun; the latter being defined through the visible limb. - The method is therefore subject to this serious drawback: that the parallax depends upon the measured differ-. ence between arcs which may be from thirty to fifty times as great as the parallax itself, the measures being made in different parts of the earth.

The equations of condition given by the American photo graphs of 1874 are found in Part I of Observations of the Transit of Venus, December 9, 1874; Washington, Government Printing Office, 1880. A preliminary solution of these equa- tions, the only one, however, to which they have yet been sub- jected, was published by D. P. TODD, in the *American Journal of Science* for June, 1881. (Vol. XXI, page 490.)

The photographs of 1882 have been completely worked up by Professor HARKNESS, and the results are found in the Report of the Superintendent of the Naval Observatory for 1889. The equations derived from the German heliometer measures, with a preliminary discussion of their results, are officially published by Dr. AUWERS, in the *Bericht über die deutschen Beobachtungen*, V, p. 710.

The separate results for the parallax, with the probable errors assigned by the investigators, are as follows:

			$w.$	w'
1874: Photographic distances,	$\pi =$	8.888 ± 0.040	6	1
Position angles,		8.873 ± 0.060	3	3
Measures with heliometer,		8.876 ± 0.042	5	5
1882: Photographic distances,		8.847 ± 0.012	64	6
Position angles,		8.772 ± 0.050	4	4
Measures with heliometer,		8.879 ± 0.025	16	10

Under w is given a system of weights proportionally deter- mined from the probable errors as assigned. Using this sys- tem, the mean result is—

$$\pi = 8''.854 \pm ''.016$$

I conceive, however, that these relative weights do not correspond to the actual precision of the measures. The very small probable error assigned by Prof. HARKNESS to the result of the photographic distances of 1882 does not include the probable error of the angular value of the unit of distance on the plate, which may arise from a number of sources, including the possible deviation of the mirror of the instrument from a perfect plane. From this error the position angles are entirely free. I have, therefore, assigned another set of weights, w', which seem to me to correspond more nearly to the facts. The result of this system is—

$$\pi = 8''.857 \pm ''.016$$

This mean error is derived from the individual discordances, and not from comparisons with the values of the parallax otherwise determined. As there may be a fortuitous agreement among the separate values, another estimate may be made on the basis of the total mean error derived by AUWERS, which includes all known sources of error. He finds $\epsilon = \pm ''.032$ for the combined heliometer results, to which I have assigned weight 15. Hence, for the total weight 29, we have—

$$\epsilon = \pm 0''.023$$

The deviation of the above result from the mean of all the other good ones is worthy of special attention. The deviation is more than three times its mean error, and therefore between four and five times its probable error. We must therefore accept one of two conclusions, either the probable errors have been considerably underestimated, or the method is affected with some undiscoverable source of systematic error, which makes it tend to give too large a result. The close accordance of the six separate results, of which only a single one deviates from the adopted mean by more than its probable error, and that by only a little more, would give color to the view that the error is a systematic one, and that through some unknown cause Venus is always measured too low relatively from the center of the Sun. I can not, however, think of any such cause.

If we determine the mean error from the deviations of the separate results from what we know, in other ways, to be nearly the most probable value of the parallax, namely 8''.80, we have—

$$\text{Mean errror to weight 1; } \pm .148''$$
$$\text{Mean error of result } \quad \pm .029$$

Solar parallax from observed contacts during transits of Venus.

74. The contact observations of 1761 and 1769 are discussed in *Astronomical Papers*, Vol. III. I have also made a complete discussion of those of 1874 and 1882, which, at the date of writing, is unpublished. The separate results from each contact follow.

In the case of the second contacts of 1874 and 1882 it was found necessary to divide the observations into two classes: those of mean or true contact, and those of the formation of the thread of light. In the case of the third contact no such division was necessary, as the observations could generally be referred to the same mean phase. The mean error which follows each result is derived from the discordance of the separate observations.

Values of the solar parallax from observed contacts of the limb of Venus with that of the Sun.

1761,	III;	$\pi = 8.78 \pm .12$;	$w. =$	8
	IV;	$8.75 \pm .20$		3
1769,	I;	$9.04 \pm .17$		4
	II;	$8.55 \pm .13$		7
	III;	$8.72 \pm .09$		14
	IV;	$9.01 \pm .12$		8
1874,	I;	$8.95 \pm .24$		2
	II; M;	$8.78 \pm .061$		30
	II; L;	$8.75 \pm .10$		11
	III;	$8.76 \pm .045$		57
	IV;	$8.74 \pm .09$		14
1882,	I;	$8.93 \pm .15$		5
	II; M;	$8.76 \pm .042$		64
	II; L;	$8.72 \pm .072$		22
	III;	$8.88 \pm .042$		64
	IV;	$9.07 \pm .12$		8

The weights assigned are determined by these mean errors, taken on such a scale that unity is the weight for mean error $\pm$ ''.336. The mean result of the whole series is

$$\pi = 8''.797 \pm ''.023$$

This mean error is that resulting from the deviations of the sixteen separate results from the general mean, which give for the mean error corresponding to weight unity,

$$\varepsilon_1 = \pm ''.42.$$

The excess of this mean error over that determined from the equations themselves shows that the general discordance of the several contacts is somewhat greater than would be inferred from the individual discordances of the contacts *inter se*. This is what we should expect from constant errors in the determinations of parallax from each separate contact. I conceive, however, that such constant errors are not likely to be large; and we can not conceive that contact observations in general are subject to any constant error tending to make the parallax derived from them always too great or too small. I conclude, therefore, that the mean error determined from the totality of the results may be regarded as real.

It will be interesting to compare the separate results of internal and external contacts. They are

From internal contacts; $\pi = 8.776 \pm .023$
From external contacts; $\pi = 8.908 \pm .06$

These mean errors are those derived from the concluded results and they show that the external contacts are relatively more discordant in proportion to the weights assigned than are the internal ones. If we consider this discordance to indicate a larger mean error, and therefore assign a proportionally smaller weight to the results of external contact, we have, for the concluded result,

$$\pi = 8''.791 \pm ''.022$$

As these two hypotheses seem about equally probable, I shall adopt the mean result,

$$\pi = 8''.794$$

Solar parallax from the observed constant of aberration and measured velocity of light.

75. The question of the soundness of the proposition that the aberration is equal to the quotient of the velocity of the Earth in its orbit by the velocity of light is too broad a one to be discussed here. I can only remark that its simplicity and its general accord with all optical phenomena are such that it seems to me it should be accepted, in the absence of evidence against it.

In *Astronomical Papers*, Vol. II, page 202, I have given the following determinations of the velocity of light in vacuo by MICHELSON and myself, expressed in kilometers per second:

MICHELSON at Naval Academy in 1879	299910
MICHELSON at Cleveland, 1882	299853
NEWCOMB at Washington, 1882, using only results supposed to be nearly free from constant errors .	299860
NEWCOMB, including all determinations	299810

I have concluded,

Velocity of light in vacuo, = 299860 ± 30 k. m.

Taking as the equatorial radius of the Earth 6378.2 k. m. (CLARK), the following table shows the values of the constant of aberration corresponding to admissible values of the solar parallax when this determination of the velocity of light is accepted.

Ab. = 20.46″	π = 8.8076″
20.47	8.8033
20.48	8.7990
20.49	8.7946
20.50	8.7903
20.51	8.7859
20.52	8.7816
20.53	8.7773
20.54	8.7730

We thus have for the values of the solar parallax resulting from the two values of the constant of aberration already derived:

From Pulkowa determinations;   Ab. $= 20.493$; $\pi = 8.793$
From miscellaneous determinations; Ab. $= 20.463$; $\pi = 8.806$

Solar parallax from the parallactic inequality of the Moon.

76. I have derived a value of the parallactic inequality of the Moon from the meridian observations made at Greenwich and Washington since 1862. The determination of this inequality is peculiarly liable to systematic error, owing to the fact that observations have to be made on one limb of the Moon when the inequality is positive, and on the other limb when it is negative. Hence, if we determine the inequality by the comparison of its extreme observed effects on the Moon's longitude or Right Ascension, any error in the adopted semidiameter of the Moon will affect the result by its full amount.

It does not seem practicable to make a reliable determination of the Moon's diameter, because it will necessarily be made near the time of full Moon, when the illumination of the extreme limb is less intense than near the quadratures, and when some portions of the limb that might be visible if it were illuminated by a perpendicular Sun will be thrown into shadow by the horizontal one. For these reasons it may be expected that the parallactic inequality determined by using observed semidiameters of the Moon will be too large. I have therefore adopted the plan of determining the inequality from each limb separately. To show in regular progression the errors depending on the elongation from the Sun, I have classified the residuals of observations according to the hour of mean time at which the Moon passed the meridian; and formed equations of condition containing two unknown quantities, the one a constant correction depending on the semidiameter, personal equation, etc., and the other the parallactic inequality. The question is further complicated by the fact that the majority of observations near are quadratures made during daylight, when it is to be expected that the illumination of the atmosphere will diminish the irradiation, and thus lead to a smaller apparent semidiameter. I have therefore sought to determine for the two observatories, by a comparison of the observations, the correction to be applied in order to reduce observations made during daylight or twilight to what they would have been had the sky not been illuminated. The reduction was smaller than I had expected, and somewhat doubtful; I have assigned proportionally less weight to those observations where it was necessary. The following are the equations of condition thus formed. The unknown quantities are—

x, a constant, depending on the semidiameter, personal equation, etc.;

y, the correction to the parallactic inequality of the Moon after reduction to the value $8''.848$ of the solar parallax.

GREENWICH.

Limb I.

h		$''$		
4.6;	$x + 0.93\,y$	$= -0.53$;	wt.	0.2
5.6	0.99	-0.72		0.6
6.5	0.99	-0.41		1
7.5	0.92	-0.59		1
8.5	0.79	-0.54		1
9.5	0.61	-0.13		1
10.5	0.38	-0.09		1
11.5	0.13	-0.06		1

Limb II.

		$''$		
12.5;	$x' - 0.13\,y$	$= +0.20$;	wt.	1
13.5	-0.38	$+0.16$		1
14.5	-0.61	$+0.28$		1
15.5	-0.79	$+0.54$		1
16.5	-0.92	-0.11		1
17.5	-0.99	-0.02		1
18.4	-0.99	$+0.44$		0.5
19.4	-0.93	$+1.21$		0.2

WASHINGTON.

. Limb I.

$$\begin{array}{llll}
h & & '' & \\
4.6; & x + 0.93\,y & = -1.62; & wt. = 0.2 \\
5.6 & 0.99 & -1.26 & 0.4 \\
6.5 & 0.99 & -0.85 & 1 \\
7.5 & 0.92 & -0.64 & 1 \\
8.5 & 0.79 & -0.71 & 1 \\
9.5 & 0.61 & -0.71 & 1 \\
10.5 & 0.38 & -0.48 & 1 \\
11.5 & 0.13 & -0.23 & 1
\end{array}$$

Limb II.

$$\begin{array}{llll}
 & & '' & \\
12.5; & x' - 0.13\,y & = +0.41; & wt. = 1 \\
13.5 & -0.38 & 0.43 & 1 \\
14.5 & -0.61 & 0.52 & 1 \\
15.5 & -0.79 & 0.40 & 1 \\
16.5 & -0.92 & 0.72 & 1 \\
17.5 & -0.99 & 0.96 & 0.5 \\
18.4 & -0.99 & 1.32 & 0.3 \\
19.4 & -0.93 & 1.50 & 0.1
\end{array}$$

With these equations we have our choice to determine the parallactic inequality by assigning a value to the semidiameter, or to eliminate the semidiameter from the normal equations. In each case the equations give the following expressions for y:

$$\begin{array}{ll}
 & '' \qquad '' \\
\text{Greenwich:} \quad \text{Limb I;} & y = -0.55 - 1.23\,x \\
\text{\textquotedblleft} \qquad\qquad \text{\textquotedblleft} \quad \text{II;} & \quad\;\; -0.28 + 1.23\,x' \\
\\
\text{Washington: Limb I;} & y = -0.99 - 1.23\,x \\
\text{\textquotedblleft} \qquad\qquad \text{\textquotedblleft} \quad \text{II;} & \quad\;\; -0.88 + 1.29\,x'
\end{array}$$

If we choose to utilize the observed diameters we have the following results:

From 66 transits of the Moon's diameter observed at Greenwich;

$$x - x' = -0''.64$$

From 33 transits observed at Washington:

$$x - x' = - 1''.12$$

We should thus have,

From Greenwich observations, $y = - 0.02$
From Washington observations, $y = - 0.23$

If, on the other hand, we eliminate x from each pair of normal equations, the final results for y will be

			$''$	$''$	$''$	wt.
Greenwich:	Limb	I;	$0.64\,y = - 0.45;$	$y = - 0.70 \pm 0.16$		6
"	"	II;	$0.64\,y = 0.00;$	$y = 0.00 \pm 0.36$		2
Washington:	Limb	I;	$0.64\,y = - 0.52;$	$y = - 0.81 \pm 0.16$		6
"	"	II;	$0.53\,y = - 0.32;$	$y = - 0.60 \pm 0.27$		3

The weighted mean of these results is

$$y = - 0''.64 \pm 0''.12$$

The resulting value of the solar parallax is

$$\pi = 8''.802 \pm 0''.008$$

A very careful determination of the solar parallax was made from the same theory by Dr. BATTERMAN, by means of occultations, and the result is discussed very fully in the publications of the Berlin Observatory. Dr. BATTERMAN'S definitive result is

$$\pi = 8''.794 \pm ''.016$$

I have slightly revised this result, by applying a correction to the coefficient for the parallax adopted by Dr. BATTERMAN, with the result

$$\pi = 8''.789 \pm ''.016$$

Accepting this result, and combining it with that already found from meridian observations, the parallax from this method will finally come out

$$\pi = 8''.799 \pm ''.007$$

This mean error may be regarded as belonging to the doubtful class.

While this work is passing through the press there appears an important paper by FRANZ of Königsberg,* giving the value of the parallactic equation derived from observations on the lunar crater *Mösting A*. The correction to HANSEN'S coefficient is found to be

$$- 2''.10 \pm 0''.30$$

The corresponding result for the solar parallax is

$$8''.767 \pm 0''.021$$

We may combine the three results for the solar parallax thus:

Greenwich and Washington meridian observations	$\pi = 8.802$;	$w = 5$
BATTERMANN from occultations.	8.789;	2
FRANZ from crater *Mösting A*	8.767;	1
Mean.	8.794 $\pm$ ''.008	

Solar parallax from observations on minor planets with the heliometer.

77. The fact that the determination of the parallaxes of the small planets by comparison with neighboring stars is free from the grave uncertainty attaching to similar observations of Venus and Mars, owing to the absence of a sensible disk, was long since pointed out by Dr. GALLE. In 1875 he published a discussion of observations on Flora, made at nine northern observatories, and at the Cape, Cordoba, and Melbourne in the Southern hemisphere.† The result was

$$\pi = 8''.873.$$

An examination of the residuals of the several observatories shows that in the case of at least one of the Southern observatories there is a systematic difference of a considerable fraction

---

* Astronomische Nachrichten, Vol. 136, S. 354.

† Ueber eine Bestimmung der Sonnen-Parallaxe aus correspondirenden Beobachtungen des Planeten Flora, in October und November 1873. Breslau, Maruschke & Berendt, 1875.

of a second. This fact seems to prevent our assigning any appreciable weight to the final result.

In 1874, GILL, at Mauritius, made heliometer observations of Juno, east and west of the meridian, with the same object. The result was 8″.765, or 8″.815 when a discordant observation was rejected. In this connection, only an allusion is necessary to GILL'S expedition to Ascension in 1877, made for the purpose of applying the method to Mars at the opposition of that year.

Shortly afterwards GILL published in the first volume of *The Observatory* a very exhaustive discussion of the methods of determining the solar parallax, in which he showed that heliometer observations of the minor planets, made either at a single station not too far from the equator, or at two stations in different hemispheres, afforded a method of measuring the parallax more precise than any before applied.

Ten years elapsed before the plan was put into operation. Then, in 1889 and 1890, a concerted system of observations was made on the three minor planets, Victoria, Iris, and Sappho, at a number of observatories in both hemispheres. The observations relating to Victoria were carried out most thoroughly, in that a very careful triangulation of the stars of comparison *inter se* was made at the observatories which took part in the measures. The tabular data for the reductions were supplied by the office of the *Berliner Jahrbuch*, and the reductions and discussion were made by GILL himself for Victoria and Sappho, and by Dr. ELKIN, on GILL'S plans, for Iris. The three results, as communicated in advance of their complete official publication, are

$$\text{From Victoria: } \pi = 8.800'' \text{ p. e. } \pm 0.006''$$
$$\text{Iris: } \quad 8.825 \text{ p. e. } \pm 0.008$$
$$\text{Sappho: } \quad 8.796 \text{ p. e. } \pm 0.012$$

I assign the respective weights 4, 2, and 1, thus obtaining, as the final result of this method,

$$\pi = 8''.807 \pm 0''.006$$

I have included in a separate category GILL'S determination by Mars, at Ascension, in 1877, as published by the

Royal Astronomical Society (*Memoirs Royal Astronomical Society*, Vol. XLVI), for the reason that, owing to the disk of Mars, and its reddish color, determinations made on it are liable to errors peculiar to that planet, or at least different from those which might come in in the case of the small planets.

Remarks on determinations of the parallax which are not used in the present discussion.

78. In the preceding discussion are given the results of every modern method of determining the solar parallax with which I am acquainted, except meridian and equatorial observations on Mars. I have not used any of the results derived from this source, owing to their large probable error, and the suspicion of systematic error to which they are open. One of these causes of error is to be found in the red color of Mars. This cause will be pointed out and discussed very fully in a subsequent section. Its effect would be to make the observed parallax too large. Since, as a matter of fact, all the determinations of Mars by meridian observations have given a larger parallax than the generality of other methods, color seems to be given to this suspicion. Apart from this, the setting of the threads of a meridian circle upon the apparent disk of Mars involves a visual estimate not comparable with that of the bisection of the image of a star by the threads. Hence, there is a chance of systematic personal error arising from this source. The observations generally exhibit large discordances, which may be attributed to one or the other of these causes.

It may be objected to the inclusion of GILL'S Ascension result that it should be rejected for the same reason, since the color of the planet would affect heliometer observations and meridian observations equally. I have, however, considered it free from the objection in question, for two reasons. In the first place, the result is not too large, but is, on the contrary, the smallest of all the accurate measures. The principle that when a result is open to a strong suspicion of being affected by a cause which would cause it to deviate in one direction, it is logical to conclude *a posteriori* that the cause has not acted if the deviation is found to be in the other direction, may not be a perfectly sound one, but I have nevertheless acted upon it. In the next place GILL himself, as a part of his discussion, compared the observations when Mars was at different altitudes, in order to determine whether the action of such a cause was indicated, and found a negative result.

In 1890 an unsuccessful attempt was made, at the writer's request, by Dr. W. L. ELKIN, to measure the effect in question, by placing a refracting prism of very small angle over one of the halves of a heliometer objective, and measuring the refraction thus produced. It was supposed that the dispersing action of the prism would represent that of the atmosphere, greatly magnified. The failure arose from the result that the apparent mean refraction of the star produced by the prism proved to be a function of the star's magnitude, ranging from 748″.79 for a star of magnitude 2.55 to 751″.61 for a star of magnitude 6.95. The reason seemed to be that too powerful a prism was used, so that the spectrum was quite sensible; then, in the case of faint stars, the red portion of the spectrum was invisible, so that the apparent mean refraction was greater than in the case of the brighter stars. The mean of the observed displacements of Mars was 748″.61, so that it was always less for Mars than for the stars.*

An investigation of the question whether the same effect is noticeable in meridian observations fails to show any relation between the brightness of a star and its refraction. But this does not disprove the relation between the refraction and the color of a star.

On the whole it seems to me that, at least in the case of Mars, we have here a cause so mixed up with personal error in making the observations that the objective and subjective effects can not be completely separated.

---

* *Astronomical Journal*, Vol. 10, page 97.

CHAPTER VIII

DISCUSSION OF RESULTS FOR THE SOLAR PARALLAX AND THE MASSES OF THE FOUR INNER PLANETS.

79. We have, in what precedes, found or collected nine separate values of the parallax of the Sun, by methods of which seven may be regarded as completely distinct, in the sense that no one source of error is common to any two. Of these seven the two most nearly associated are those which utilize transits of Venus. These are similar only in the sense of resting upon a determination of the relative parallax of Venus and the Sun during the time of a transit. But the only common elements which enter into the determination are the ratio of the distances of the Sun and Venus, which is determined with such certainty that we can not regard it as subject to error. The methods of determining the parallax in the two cases are.completely distinct from the beginning, there being, I conceive, no common source of error affecting an observation of contact of limbs and one of a distance measured from the center of the Sun while Venus is in transit.

I have classified as if they were independent the values of the parallax which follow from the Pulkowa determinations of the constant of aberration, and those which follow from all other determinations. Of course whatever doubts may affect the theory of the assumed relation between the constant of aberration and the velocity of light will equally affect both determinations. I do not, however, conceive that there is any source of error which can affect both the Pulkowa determinations of the aberration and those made elsewhere. The two could have been combined so as to give a single result of the method; but as the two values of the constant differ by more than we should expect them to from their probable errors, I have kept them separate, partly not to give a false appearance of agreement of results, and partly to facilitate the inception of any future investigation on the subject.

I have also separated the result of GILL's observations on Mars, at Ascension, in 1877, from the determinations made by the same method on the minor planets, because, owing to the color and disk of Mars, the two results may be affected by very different systematic errors. The only common systematic error which seems likely to affect them is that arising from the color of the object, which will be discussed hereafter.

Results of determinations of the solar parallax arranged in the order of magnitude.

	"	"	wt.
From the mass of the Earth resulting from the secular variations of the orbits of the four inner planets . . .	8.759	± .010	9
From GILL's observations of Mars at Ascension	8.780	± .020	2
From Pulkowa determinations of the constant of aberration.	8.793	± .0046	40
From observations of contacts during transits of Venus	8.794	± .018	3
From the parallactic inequality of the Moon	8.794	± .007	18
From determinations of the constant of aberration made elsewhere than at Pulkowa	8.806	± .0056	28
From heliometer observations on the minor planets	8.807	± .007	20
From the lunar equation in the motion of the Earth	8.825	± .030	1
From measurements of the distance of Venus from the Sun's center during transits	8.857	± .023	2

The mean errors which follow each value are those which, from a study of the determination, it seemed likely might affect them, no allowance being made for mere possibility of systematic error. The weights assigned are convenient small integers, generally such as to make the weight unity correspond to the mean error ± 0″.30, allowance being made, how-

ever, for doubt as to what value should be assigned to the
mean error and for the different liabilities to systematic error.
The mean result is—

From all determinations; $\pi = 8.797''$
Omitting the first result; $\pi = 8.800 \pm .0038$

The last value differs from the preliminary value $8''.802$ of
Chapter V, from a change in the weights. It will be seen
that the different values are all as accordant as could be
expected, with the exception of the two extreme ones. In the
largest value we have a case the principles involved in which
have been discussed in Chapter IV.

We can not suppose the parallax to be materially greater
than $8''.800$, and may take it as probably less than this. Thus
the absolute error of the results of measures of Venus on the
face of the Sun may be considered as about $0''.06$ or $0''.07$,
which is four times the computed probable error. The probability against this, even in the case of one result out of eight
or nine, is so small that we must either regard the method as
being affected by some systematic error, or as affected by
an objective probable error larger than that assigned. It
seems to me the latter view is not untenable, in view of the
very wide range of the possibilities of error which might affect
a series of observations with a heliometer exposed to the Sun's
rays during a period limited to a few hours.

Again, in the photographic measures, the value of a second
of arc in length on the photographic plate enters as a somewhat uncertain element. In this connection it is to be
remarked that the measures of position angle on the photographic plates, which are not affected with this uncertainty,
although their probable error is quite considerable, give a
value of the solar parallax much smaller than the measures of
distance.

Much more embarrassing is the value which results from the
mass of the Earth. We here meet in another aspect the same
deviation which we encountered in determining the mass of
the Earth from the secular variations, and on which we post-
poned a conclusion (§ 64). This determination rests very

largely on the motion of the node of Venus, as determined
from the transits of 1761 and 1769. It is true that results of
meridian observations are combined with them; but no explanation is thus afforded of the difficulty, because the results of
these observations agree with those of the transits (v. §39).
What adds to the embarrassment and prevents us from wholly
discarding the suspicion that some disturbing cause has acted
on the motion of Venus, or that some theoretical error has
crept into the work, is that, of all the determinations of the
solar parallax this is the one which seems the most free from
doubt arising from possible undiscovered sources of error. It
is, as we shall presently see, really entitled to twice the relative
weight assigned it. As, however, the determination rests
mainly on the motion of the node of Venus, and this again
mainly rests on the observations of the older transits, I have
made a reexamination of the results of these transits with a
view of reaching a more exact estimate of the sources of error
and the magnitude of the mean error. In this re-examination
I have regarded the Sun's parallax as a known quantity equal
to 8″.798, and then obtained the results of the old observations
of the transits on the supposition that the only quantities to
be determined were the corrections to the relative heliocentric
positions of Venus and the Earth.

Rediscussion of the motion of the node of Venus.

80. In discussing the observations of 1761 and 1769 (*Astronomical Papers*, Vol. II, Part V), I introduced a quantity
expressive of the error in the observed time of contact arising
from imperfections of the telescope and atmospheric absorp-
tion and dispersion. The constants on which these errors
depend are represented by symbols k_2 and k_3. As I have
worked up the observations, the ultimate result of each
observation of contact is the value of an unknown quantity,
δc, which, were there no imperfections of vision and were the
radii of the Sun and Venus accurately known, would represent
the correction to the tabular distance of centers. As a matter
of fact, however, we are to consider δc as equal to this correction increased by a rather complex combination of quantities
depending on the errors of the assumed semidiameters of

Venus and the Sun, and the thickness of the thread of light
when it first became visible at second contact, or vanished at
third contact. The observations must be so combined as to
eliminate these quantities. What I have done is to represent
the undiscoverable minute correction to δc thus arising by
the symbol z_2 for second contact, and z_3 for third contact. In
the present re-examination the absolute terms are reduced to
the parallax $8''.798$ by putting $\delta \pi_0 = -''.05$ and $\pi' = -''.025$
in the final equations of the original paper. After each result
is given the mean error with which it is affected, as determined by the investigation in question. When thus treated,
the equations which I have given on pages 391–398 of the
paper referred to give the following normal equations for δc,
the indeterminates k_2 and k_3 being retained as such in order to
show their final effect on the result.

$$1761. \quad \text{II}; \quad 8.5 \;\; \delta c = + 0.76 - 18.5\,k_2 \pm 0.78$$
$$\text{III}; \quad 41.7 \;\; \delta c = - 2.81 - 19.2\,k_3 \pm 1.30$$

$$1769. \quad \text{II}; \quad 44.8 \;\; \delta c = - 8.00 - 104.1\,k_2 \pm 1.95$$
$$\text{III}; \quad 12.1 \;\; \delta c = + 0.31 - 16.0\,k_3 \pm 0.70$$

In order to vary the proceeding as much as possible from
that of the former investigation, I now express δc in terms of
$\delta \lambda$ and $\delta \beta$, which, for the time being, I take as the corrections
to the heliocentric longitude and latitude of Venus referred
to the Earth, and these again in terms of δv and $\sin i \delta \theta$,
which latter, for brevity, I call u. The first transformation is
made with the coefficients of p. 71, where we have put x and
$- y$ for $\delta \lambda$ and $\delta \beta$, and the last by the equations

$$\delta \lambda = \delta v + 0.06\, u$$
$$\delta \beta = u - 0.06\, v$$

Putting u_1 for the value of u in 1765, we have, in consequence
of the known change in the motion of the node,

$$\text{In } 1761; \quad u = u_1 + 0.11$$
$$\text{In } 1769; \quad u = u_1 - 0.11$$

We thus have the four equations which follow for determining δv and u_1, the former being supposed the same at the times of the two transits.

$$- .84 \, \delta v - .55 \, u_1 + z_2 = + 0\overset{''}{.}15 - 2.2 \, k_2 \pm 0\overset{''}{.}09$$
$$+ .73 \qquad - .09 \quad + z_3 = + 0.01 - 0.5 \, k_3 \pm 0.03$$
$$- .69 \qquad + .73 \quad + z_2 = - 0.10 - 2.3 \, k_2 \pm 0.04$$
$$+ .81 \qquad + .60 \quad + z_3 = + 0.10 - 1.3 \, k_3 \pm 0.06$$

Eliminating z_2 and z_3 by subtracting the first equation from the third, and the second from the fourth, we have—

$$.15 \, \delta v + 1.28 \, u_1 = - 0\overset{''}{.}25 - 0\overset{''}{.}1 \, k_2 \pm 0.10$$
$$.08 \, \delta v + 1.29 \, u_1 = + 0.09 - 0.8 \, k_3 \pm 0.07$$

We thus have for u_1 the value

$$u_1 = - 0''.04 - 0.08 \, \delta v - 0.03 \, k_2 - 0.36 \, k_3 \pm 0''.05$$

δv can not be determined independently of z_2 and z_3. Assuming these quantities to be equal, we have already found it to be only $0''.02$, and may therefore, to determine its probable effect upon the result by assigning to it the value

$$\delta v = 0''.00 \pm 0''.22$$

In the former paper I have found for k_2 and k_3 the values

$$k_2 = + 0\overset{''}{.}040 \pm 0\overset{''}{.}040$$
$$k_3 = - 0.034 \pm 0.040$$

A preliminary correction of $+ 2''.02$ having been applied to the tabular orbital latitude, we have, for the epoch 1765.5,

$$\sin i \delta \theta = + 1''.99 \pm 0''.06$$

Combining this result with that of the transits of 1874 and 1882, we have the following results, which are compared with those of meridian observations:

Transits of Venus alone	$\sin i \, D_t \, \delta\theta =$	$- 2\overset{''}{.}82$
Meridian observations alone	"	$- 2.45$
Combined solution	"	$- 2.71$
Adjusted with other results (§ 46) . . .	"	$- 2.73$
Adopted	"	$- 2.77$

The adopted result is the one which seems the most probable. For the final probable error we are to include that of the precession and of the Sun's longitudes at the two epochs. We may estimate the combined value of these at $\pm 1''$, corresponding to an error of $0''.06$ in $\sin i \, D_t \, \delta\theta$. Thus we have

$$\sin i \, D_t \, \delta\theta = - 2''.77 \pm ''.084$$

I conceive this mean error to be as real as any that can be determined in astronomy. This conviction rests upon the fact (1) that the systematic errors affecting the four contacts are shown to be small by the general minuteness of the four values of δc; (2) that whatever systematic errors may affect the formation or disappearance of the thread of light are almost completely eliminated from the mean of the transits of 1761 and 1769 by the method in which the observations have been combined. The accordance of the observations of external contact made at the same transits strengthens this view.

The equation thus derived takes the place of the sixth equation of § 63 and should have twice the weight there assigned. As the mass of the Earth determined by the secular variations rests mainly on this equation, I shall first consider it alone. Expressing the theoretical secular variation of $\sin i\delta\theta$ in terms of the above observed value, we find that the observed motion of the node of Venus gives the equation

$$0''.26 \; \nu - 29''.2 \; \nu' - 43''.2 \; \nu'' = + 0''.48 \pm 0''.084 \qquad (a)$$

which gives for ν'' the value

$$\nu'' = - 0.0111 + 0.006 \; \nu - 0.676 \; \nu' \pm .0019$$

The value of the solar parallax for $\nu'' = 0$ is $8''.811$. Hence, for the value expressed in terms of the corrections to the assumed masses of Venus and Mercury, this equation gives

$$\pi = 8''.778 + 0''.020 \; \nu - 1''.98 \; \nu'$$

We have found from the periodic perturbations

$$\nu = - \overset{''}{0.055} \pm .25$$
$$\nu' = + 0.0080 \pm .0025$$

Whence,

$$\nu'' = -\ 0.0168 \pm .0029$$
$$\pi =\ \ \ \ 8.762\ \ \pm .0086$$

This result of observation, errors and unknown actions aside, I can not suppose to be affected by any other mean error than that here assigned.

We have now to consider how far this result may be reconciled with the others by changes in the masses of Mercury and Venus. No admissible change in the former could greatly affect the result. The question then arises whether the discrepancy may not be due to an error in the concluded mass of Venus. In making so large a change in this element, we meet with insuperable difficulties. The observed motion of the ecliptic, which is a fairly well-determined quantity, indicates a still further increase of this mass. We may put this difficulty in another form. The observed motion of the node of Venus is a relative one, consisting in the combined effect of the motion of the ecliptic around an axis at right angles to the node of Venus, and an absolute motion of the orbit of Venus around nearly the same axis. This motion of the ecliptic depends mainly on the mass of Venus; the absolute motion of the orbit of Venus mainly on that of the Earth. If, now, we determine the motion of the ecliptic from observation, we shall find that the relative motion of the orbit of Venus still unaccounted for is yet greater than we have supposed it to be, and should therefore find a yet smaller mass of the Earth than that heretofore concluded.

The determination of the mass of Venus already made from observations of the Sun and Mercury seems to admit of no doubt. We can not conceive that the mean of fifteen determinations, made during one hundred and thirty years, at different observatories, which determinations are so separated as to be entirely independent of each other, can be affected by any considerable common error. The entire accordance of the result thus reached from the periodic perturbations produced by Venus with that from a combination of all the secular variations, as shown in Chapter VI, strengthens the result yet further. Unknown actions and possible defects of theory aside, it seems to me that the value of the solar parallax derived from this discussion is less open to doubt from any known cause than any determination that can be made.

Possible systematic errors in determinations of the parallax.

81. We have now to return to the other values, in order to see to what extent they may be affected by systematic error. I have already excused myself from discussing the validity of the assumed relation between the constant of aberration and the velocity of light, because there is nothing valuable to be said on the subject, and have alluded to the possible sources of systematic error in the Pulkowa determinations of aberration. It is worthy of attention here that the very best of these determinations, that of NYRÉN with the prime vertical transit, in respect to the care with which it was made, and the general accordance of the entire work throughout, gives a result most accordant with that under consideration. In fact, to the value 8″.77 of the solar parallax corresponds the value 20″.55 of the constant of aberration, which is larger by only 0″.02 than the result of NYRÉN's best determinations.

As for miscellaneous determinations of the constant, it is to be remembered that the corrections applied to a part of the separate values on account of the Chandlerian inequality of latitude are somewhat doubtful, and the general mean may have been affected by a few hundredths of a second in consequence. It is not, however, possible to determine the amount of the correction, except by an exhaustive rediscussion of the whole of the original observations, and even then the result would still be doubtful.

Next in the order of weight we have the results of measures on the minor planets with the heliometer, on GILL's plan. I have already remarked upon the possible error in such observations arising from the probable difference of color between the planet and the star. A hypothetical estimate of the amount of this error is worth attempting. Let us assume that in the case of a minor planet the mean of the visible spectrum corresponds to the line D, and that in the case of a star the same mean is halfway between the lines D and E.

The index of refraction of air has been determined independently by KETTLER and LORENTZ for the different rays. The mean of their results for the rays D and E is

$$\text{For D, } n = 1.000\ 2929$$
$$\text{For E, } n = 1.000\ 2940$$

These results are accordant in giving a dispersion between these two lines equal to about .0037 of the total refraction. We have hypothetically taken the extreme possible difference between planet and star to be one-half of this. At an altitude of 45°, where the refraction is about 60″, the error would be 0″.11. At an altitude of 30° the error would be 0″.20. We are thus led to the noteworthy conclusion:

If the difference between the spectra of a minor planet and a comparison star is such that the means of their respective visible spectra, or the apparent amounts of their respective refractions, differ by one-tenth of the space between D and E, an error of 0″.02 or 0″.03 may be produced in the apparent parallax of the planet.

The question thus arising may be readily settled by measures with the heliometer. The distances of pairs of stars differing as widely as possible in color should be measured at different altitudes, when one is nearly above or below the other, in order to see what difference of refraction depending on the color is indicated. A colored double star, such as β Cygni, might also be used for the same purpose.

The minor planets are of different colors. I am not aware of any evidence that Victoria or Sappho differ in color from the average of the stars, but I believe that Iris is somewhat yellow, or reddish. Now, in this connection, it is a significant fact that the parallax found from observations of Iris, 8″.825, is the largest by GILL's method.

I have already remarked that the value of the solar parallax derived from the parallactic equation of the Moon is one of which the probable mean error is subject to uncertainty. While it is true that the value may be smaller than that we have assigned, we must also admit that it may be much larger.

The probable error of the determination by the lunar equation of the Earth is larger than that of any other method. At the same time I do not think that it is liable to systematic error, and we must therefore regard the mean error assigned as real.

Results for the solar parallax after making allowance for probable systematic errors.

82. Let us now see whether we can reach a satisfactory result by making a liberal allowance for the more or less probable sources of systematic error just pointed out. The modifications we make in the weights formerly assigned are these: We reduce the weight of GILL's Ascension result to one-half, owing to the uncertainty arising from the color of the planet Mars. We retain the Pulkowa determinations of the constant of aberration with their full weight, but reduce the weight of the miscellaneous determinations. In the case of the parallactic inequality, we reduce the weight for the reasons already given. We omit Iris from the determination from the minor planets. We also reduce to one-half its former value the relative weight assigned to measures of Venus on the Sun, on the theory that the actual mean error must be larger than that given by the discordance of results. Our combination will then be as follows:

		$''$	wt.
From the motion of the node of Venus	$\pi =$	8.768	10
From GILL's Ascension observations		8.780	1
From the Pulkowa constant of aberration . . .		8.793	40
From contacts of Venus with the Sun's limb . .		8.794	3
From heliometer observations on Victoria and Sappho		8.799	5
From the parallactic inequality of the Moon . .		8.794	10
From miscellaneous determinations of the constant of aberration		8.806	10
From the lunar inequality in the motion of the Earth		8.818	1
From measures on Venus in transit		8.857	1

Mean result, ignoring the first; $8''.7905 \pm .0045$

This mean result still differs from that given by the motion of the node of Venus by more than five times the probable error of the latter, and is yet farther from the combined result of all the secular variations, so that no reconciliation is brought about.

The embarrassing question which now meets us is whether we have here some unknown cause of difference, or whether the discrepancy arises from an accidental accumulation of fortuitous errors in the separate determinations. We have already discussed the former hypothesis, and have been unable to find any reasonably probable cause of abnormal action. The motion of the planes of the orbits is that which is least likely to deviate from theory, because it is independent of all forms of action depending upon distance from the Sun, or upon the velocity of the planet.

An examination and comparison of all the results shows one curious feature: the unanimity with which the secular variations speak against the large value of the solar parallax, or of the mass of the Earth, as the one quantity at fault. The adopted motion of the node of Venus is sustained not only by the meridian observations, but by the external contacts at the transits of 1761 and 1769, and, weakly, by a comparison of the transits of 1874 and 1882.

If we determine the correction of the mass of the Earth from other secular variations than that of the node of Venus, by the equations of § 63, we have, after eliminating the masses of Mercury and Venus,

$$\nu'' = -\ 0.029;\ \text{p. e.} \pm .018$$

If, instead of eliminating these values, we put

$$\nu = +\ .08;\ \nu' = +\ .0080;$$

we have

$$\nu'' = -\ 0.026;\ \text{p. e.} \pm .014$$

In each case the value of the parallax is yet smaller than that found from the motion of the node of Venus. I have already remarked that the observed motion of the ecliptic indicates an increase of the mass of Venus.

The question thus takes the form, whether it is possible that the mean of the seven determinations of the solar parallax

$$\pi = 8''.797 \pm ''.0035$$

can with reasonable possibility be in error by an amount the correction of which would bring it within the range of adjustment of the other quantities.

From what has already been said of the systematic errors to which every one of the determinations may be liable, it is evident that we should have no difficulty in accepting the necessary reduction of each of the separate values. The improbability which meets us is not so much the amount of the individual errors of the determinations as the fact that seven of the eight independent determinations should all be largely in error in the same direction.* Still, under the circumstances, we must admit this possibility, and make what seems to be the best adjustment of all the results.

Definitive adjustment.

83. In making the definitive adjustment I shall proceed on the supposition that no correction is necessary to the adopted mass of Mars. I also go on the principle that no result is to be rejected on account of doubt or discordance, except when it is affected with a well-established cause of systematic error, and shows a large deviation in the direction in which this cause would act. At the same time it will be admissible to diminish the weights in special cases, on account of causes of systematic error which we know to exist, although we can not determine the directions in which they would act; and also on account of deviations so wide as to show that the probable error of the result must have been greatly underestimated. Proceeding on this plan, we might reweight the last eight results for the solar parallax, so as to get a result slightly different from $8''.797$. But I doubt whether such a reweighting would not involve an objectionable bias.

We might diminish the weight of the result given by the Pulkowa constant of aberration on the ground that no one method should have so preponderating a weight as this has. If we did so the result might be increased to $8''.800$. We

---

* For a very searching criticism of the systematic errors with which the determinations of the solar parallax may be affected, reference may be made to the first two articles by Dr. DAVID GILL, in Vol. I of *The Observatory*.

might very largely increase the relative weight assigned to the heliometer observations on Victoria and Sappho, but no admissible increase would appreciably change the result. We might also diminish the relative weight of the largely discordant result derived from measures of Venus during transit. But as, by throwing out this result altogether, we should only diminish the mean by $''.001$, it is scarcely worth while to do so. Altogether no rediscussion of the relative weights seems necessary.

On the other hand, the weight which we assign to the mean result will enter as a very important factor into the final adjustment. This is a point on which it is impossible to reach a positive numerical conclusion by any mathematical process.

If, as one extreme case, we consider that the mean error of each separate result corresponds to $\pm 0''.03$ for weight unity, we shall have a mean error of $\pm ''.0035$ for the value $8''.797$. The result will not be very different if we determine the mean error from the discordance of the eight separate results. On the other hand, if we include the deviation of the result given by the motion of the node of Venus, the mean error for weight unity will be increased to $\pm 0''.0045$. The latter is undoubtedly the most logical course, so long as we proceed on the hypothesis that the deviations of the final adjustment can all be explained as due to fortuitous errors. If we include a comparison with the results of all the secular variations we shall have a yet larger mean error. To show the result of assigning one weight or the other I shall make two solutions, A and B, in one of which a less and in the other a greater weight will be assigned.

To the value $8''.797 \pm .005$ or $\pm .007$ of the solar parallax corresponds

$$v'' = - 0.049 \pm .0016 \text{ or } \pm .0025$$

According as we assign one weight or the other to this result, we may take as the corresponding equation of condition of weight unity

or (A); $\qquad 400 v'' = - 2.0$
    (B); $\qquad 600 v'' = - 2.9$ $\qquad\qquad (a)$

The masses of Venus and Mercury, determined by methods independently of the secular variations, also enter as conditions into the adjustment. I have, however, made a revision of the preliminary adjustment given in § 64, the latter being based on the results of §§ 32-38; whereas it is better to use the definitive results of the combination used in § 46.

For the mass of Mercury the result found in § 53 by the last combination is

$$m = \frac{1 \pm 0.35}{7\,943\,000} \qquad (b)$$

The values of the denominator corresponding to the mean limits here assigned are

$$5\,890\,000 \text{ and } 12\,210\,000$$

These limits are so wide as to include all admissible results for the mass of Mercury. Moreover, we can not definitely say that the value (b) of this mass is markedly greater or less than that given by the weighted mean of all other results, since we might so weight the latter as to give a result greater or less without transcending the bounds of judicious judgment. I conceive, therefore, that we are justified in reducing the mean error to ± 0.26, which will give as the equation of condition

$$\nu = -0.055 \pm 0.25$$

and hence

$$40\,x = -0.22 \pm 1 \qquad (c)$$

When, in the normal equation for the mass of Venus, given by the observations on Mercury, we substitute the values of the secular variations found from the general combination of § 46, the result is

$$\nu' = -0.0114$$

Combining this with the result from the Sun, we have

$$\nu' = -0.0117$$

In view of the fact that the mass derived from observations of Mercury may be affected by systematic errors of the kind

shown and discussed in § 53, the mean error formerly assigned
to this result should be somewhat diminished. The result is

$$m' = \frac{1}{400\ 600}$$

From this we have

$$\nu' = +\,0.0084 \pm .0030$$

For the equation of condition of weight unity I take

$$330\ \nu' = +\,2.8 \qquad\qquad (d)$$

With these equations of condition we have to combine the
eleven equations of § 63, which we use unchanged, except that
we double the weight assigned to the sixth equation, that
derived from the motion of the node of Venus, on account of
the smaller probable error of the result of our preceding rediscussion, and use the value of the absolute term found in § 80.

If we accept the view that all the perihelia move according
to the same law of gravitation toward the Sun, namely, that
expressed by HALL's hypothesis, then the value of the quantity δ in the formula expressing the law of gravitation is so
well determined by the motions of Mercury that it becomes
legitimate to use the observed motions of the perihelia of the
other three planets as equations of condition. But since it is
not impossible that the minor planets between Mars and
Jupiter may have an appreciable influence on the motion of
the perihelion of Mars, it is a question whether we should not
exclude that motion from the equations.

The conditional equations given by the motions of the three
perihelia in question are found by comparing the results of
§§ 46, 54, and 61. They are

$$
\begin{aligned}
40\,x +\ \ \ 0\,\nu' + 20\,\nu'' &= \dotplus 1.0 \\
-14\ \ \ + 46\ \ \ \ + 0\ \ \ \ &= -0.3 \qquad (e)\\
2\ \ - 13\ \ \ + 61\ \ \ \ &= +0.7
\end{aligned}
$$

The conditional equations to be combined are the eleven
equations of § 63, the sixth of which is to have double weight,
and the six equations (a), (c), (d), and (e).

The normal equations to which we are thus led are the following, which show the results of the four combinations we may make according as we use (A) or (B) for the equation given by the mass of the Earth, and omit or include the third equation (d), which is given by the motion of the perihelion of Mars.

(*α*.) *Including the motion of the perihelion of Mars.*

$$9\,607x - \quad 7\,147v' - \quad 11\,335v'' = +\,220$$
$$-\ 7\,147 \quad +\,267\,174 \quad +\,168\,727 \quad\quad = -\,587$$
$$-\,11\,335 \quad +\,168\,727 \quad +\,406\,300 \quad\quad = -\,3388\,(A)$$
$$-\,11\,335 \quad +\,168\,727 \quad +\,606\,300 \quad\quad = -\,4328\,(B)$$

(*β*.) *Omitting the motion of the perihelion of Mars.*

$$9\,603x - \quad 7\,121v' - \quad 11\,457v'' = +\,219$$
$$-\ 7\,121 \quad +\,267\,003 \quad +\,169\,520 \quad\quad = -\,578$$
$$-\,11\,457 \quad +\,169\,520 \quad +\,402\,578 \quad\quad = -\,3431\,(A)$$
$$-\,11\,457 \quad +\,169\,520 \quad +\,602\,578 \quad\quad = -\,4371\,(B)$$

The results of the solutions in the four cases are:

	A*α*	A*β*	B*α*	B*β*
x	+ 0.0147	+ 0.0142	+ 0.0161	+ 0.0158
v	+ 0.147	+ 0.142	+ 0.161	+ 0.158
v'	+ 0.004 34	+ 0.004 60	+ 0.003 10	+ 0.003 25
v''	− 0.009 73	− 0.010 05	− 0.007 70	− 0.007 87
$1 \div m$	6 539 000	6 567 000	6 460 000	6 477 000
$1 \div m'$	408 230	408 120	408 730	408 670
π	8″.783	8″.782	8″.789	8″.788

I conceive that if the secular variations, especially the motion of the node of Venus, are not affected by any unknown cause, some mean between these should be regarded as the most probable solution. The result does not, however, bring about a satisfactory reconciliation. We still find ourselves confronted by this embarrassing dilemma: Either there is something abnormal in connection with the node of Venus, due to an unknown cause acting on the planet, to some extraordinary errors in the observations or their reduction, or to some error in the theory on which the discussion is based, or the determinations of the solar parallax are nearly all in error in one direction by amounts which are, in more than one case, quite surprising.

Possible causes of the observed discordances.

84. Two possible causes of discordance may be suggested, one of which has not been touched upon at all in the preceding chapters, and one perhaps inadequately. As to the hypothesis of non-sphericity of the Sun, considered in §56, it may be remarked that Dr. HARTZER shows that an ellipticity of the Sun sufficient to produce the observed motion of the perihelion of Mercury would cause a direct motion of $5''.1$ in the motion of the node of Venus. This would correspond to a change of $0''.30$ in the value $\sin i \, D_t \theta$ and would therefore go far toward reconciling the discrepancy. But it is easy to see that this cause would produce a secular motion of $-2''.6$ in the inclination of Mercury. We have seen that the observed motion of the inclination already exceeds the theoretical motion by $0''.38$; so that introducing the hypothesis of ellipticity of the Sun we should have a discrepancy of about $3''.0$ between theory and observation. This conclusion alone seems fatal to the theory, which otherwise has been shown to be scarcely tenable.

The other possible cause is an inequality of long period; especially one depending on the argument $13 l'' - 8 l'$ which has a period of about two hundred and forty-three years. A very simple computation shows that the coefficient of this term is only of the order of magnitude $0''.01$.

It is a curious coincidence that if we had neglected to add the mass of the Moon to that of the Earth, in computing the secular variations, the discrepancy would not have existed.

Adopted values of the doubtful quantities.

85. The practical question which has been before the writer in working out the preceding results is: What values of the constants should be used in the tables of the celestial motions of which the results of this discussion are to form the basis? Should we aim simply at getting the best agreement with observations by corrections more or less empirical to the theory? It seems to me very clear that this question should be answered in the negative. No conclusions could be drawn from future

comparisons of such tables with observations, except after reducing the tabular results to some consistent theory. The imposition of such a labor upon the future investigator is not to be thought of. Moreover, there is no certainty that the tables which would best represent past observations would also best represent future ones. Our tables must be founded on some perfectly consistent theory, as simple as possible, the elements of which shall be so chosen as best to represent the observations.

In choosing the theory and its constants we have again a certain range. If we accept the necessity of assuming the secular variations of the orbits of Mercury and Venus to be affected by the action of unknown masses of matter, then the simplest course to adopt is to construct our theory on the supposition of a planet or group of planets between Mercury and Venus.

It seems to me that the introduction of the action of such a group into astronomical tables would not be justifiable. The more I have reflected upon the subject the more strongly seems to me the evidence that no such group can exist, and, indeed, that whatever anomalies exist can not be due to the action of unknown masses of matter.

Besides, the six elements of such a group would constitute a complication in the tabular theory.

On the other hand, it did not seem to me best that we should wholly reject the possibility of some abnormal action or some defect between the assumed relations of the various quantities. What I finally decided on doing was to increase the theoretical motion of each perihelion by the same fraction of the mean motion, a course which will represent the observations without committing us to any hypothesis as to the cause of the excess of motion, though it accords with the result of HALL's hypothesis of the law of gravitation; to reject entirely the hypothesis of the action of unknown masses, and to adopt for the elements what we might call compromise values between those reached by the preceding adjustment and those which would exist if there is abnormal action. The exigency of having to prepare the tables required me to reach a conclusion on this subject before the final revision of the preceding discussion, so that the numbers used are not wholly based upon it. The conclusions I have reached are these:

Since, if there is nothing abnormal in the theory, the solar parallax is probably not much larger than 8″.780, and if there is anything abnormal it is probably as large as 8″.795 or even 8″.800, we may adopt the value 8″.790 as one which is almost certainly too large on the one hypothesis and too small on the other, and which is therefore best adapted to afford a decision of the question.

For the mass of Venus I took, as an intermediate value,

$$m' = 1 \div 408\ 000$$

For the mass of Mercury I took

$$1 \div 6,000,000$$

Actually it seems that this mass is larger than the most probable one on either hypothesis, though not without the range of easy possibility.

With these values the outstanding difference between theory and observation in the centennial motion of the node of Venus is

$$\Delta \sin i\, D_t\, \theta = 0''.25$$

If this difference arises wholly from the error of the theory, then between the transits of 1874 and 2004 the accumulated error would amount to 0″.32 in the heliocentric latitude, and about 0″.8 in the geocentric latitude. Unless an improvement is made in the method of determining the position of Venus by observation, the twentieth century must approach its end before this difference can be detected.

Bearing of future determinations on the question.

86. The following shows the influence which subsequent determinations of the principal elements will have upon our judgment as to the solution of the dilemma. The changes in the second column will, by emphasizing the discordance between the results, tend to confirm the hypothesis of an abnormal defect in the theory, while the opposite ones, in the last column, will tend to reconcile theory and observation:

Element or quantity.	Change tending to confirm the discordance between theory and observation.	Change tending to reconcile existing theory with observation.
The solar parallax.	*Increase.*	*Diminution.*
Longitude of the node of Mercury.	*Increase.*	*Diminution.*
Longitude of the node of Venus.	*Increase.*	*Diminution.*
Constant of aberration.	*Diminution.*	*Increase.*
Mass of Venus.	*Increase.*	*Diminution.*
Mass of Mercury.	*Diminution.*	*Increase.*
Secular diminution of the obliquity.	*Diminution.*	*Increase.*

. Among these constants are some the values of which can scarcely be decisively obtained except by observations continued through half a century, or even through the whole twentieth century, unless improvements are made in our present methods of observing.

The improvement of others, however, is quite within the reach of the astronomy of the present time. Among these the constant of aberration and the solar parallax have the first place. The more accurate determination of these quantities thus assumes an importance which may justify some suggestions on the subject.

The observations made on the European continent for the detection and study of the variations of latitude have been executed with such precision that we might look to them for a marked improvement in the determination of the constant of aberration, were it not for a single circumstance. In the general average few are made after midnight, while the maxima and minima of aberration occur in the morning and evening. The extension of the system into the early morning therefore seems desirable. Although these observations may scarcely equal in accuracy those made by NYRÉN, with the prime vertical transit, they have the advantage of not requiring so long a period for a complete observation. The great disadvantage of the prime vertical instrument is that unless a star culminates within a few minutes of the zenith, an hour, or even several hours, will be required for the completion of a determination, which may thus be made impossible by the 'advent of daylight. It may be remarked in this connection that the northern latitudes of the European observatories are favorable to the determination of the aberration-constant.

LOEWY's method has over all others the great advantage of being independent of the direction of the vertical. But its application, and the reduction of the observations made with it, are laborious in a high degree.

So far as practical astronomy has yet developed, the best method of directly measuring planetary parallax, and therefore the only one to be considered, is that of GILL. It therefore seems desirable that measures by this method should be continued. At the same time it is very necessary that the spectra of the small planets to be used should be carefully studied photometrically, and that the probable influence of coloration upon the measures should be investigated.

The necessity of completing the present work, and of proceeding immediately to the construction of tables founded upon the adopted elements, prevent the author's awaiting the mature judgment of astronomers upon the embarrassing questions thus raised. The regret with which he accepts this necessity is weakened by the consideration that even if the solar parallax which he has adopted requires the largest correction to which it can reasonably be supposed subject, namely, one of $-0''.015$, reducing the value of this constant to $8''.775$, the effect of the error will not be prejudicial to the astronomy of the immediate future.

More important will be the error $0''.035$ in the constant of aberration. Yet a long-continued series of observations will be necessary to establish even the existence of such an error, and should it prove detrimental in any astronomical work the evil will be easily remedied by a slight correction.

5690 N ALM——12

CHAPTER IX.

DERIVATION OF RESULTS.

Ulterior corrections to the motions of the perihelion and mean
. longitude of Mercury.

87. In §§32 and 46 we have reached three values of the correction to the tabular motion of the perihelion of Mercury. Of these the first rests on meridian observations alone, the second on the combination of meridian observations with trans-its, and the third is derived by substituting in the eliminating equations the corrections to the solar elements and their secular variations which result from observations. The three values thus reached are $-9''.54$, $-1''.01$, and $+6''.34$. The progressive divergence of these values, taken in connection with the discrepancy pointed out in §33, leads us to distrust the influence of the meridian observations upon the motion of the perihelion. Under these circumstances I deem it advisable to make such final corrections to the motions in mean longitude and mean anomaly as will best satisfy all the observed transits over the disk of the Sun. In doing this I am enabled to intro-duce the results of a preliminary discussion of the transits of 1891 and 1894. By combining the observations of these two transits with those of the older ones I derive the following values of the functions V and W defined in §31:

$$V = -1.93 - 3.03\,T$$
$$W = +1.50 + 2.04\,T$$

The preliminary theory, so far as yet investigated, gives for the values of this quantity,

$$V = -2.44 - 3.40\,T$$
$$W = +1.38 + 1.36\,T$$

178

Equating these values to the corresponding linear functions of the corrections to l, π, and their secular motions, we have the equations,

$$0.72\,\delta l + 0.28\,\delta\pi = +\overset{''}{0.12} + \overset{''}{0.68}\,T$$
$$+1.49 \quad -0.49 \quad = +0.51 + 0.37\,T$$

We find, from these equations,

$$\delta l = +\overset{''}{0.26} + \overset{''}{0.56}\,T$$
$$\delta\pi = -0.24 + 0.97\,T$$

The preliminary values to which these corrections are applicable are

$$\delta l = +\overset{''}{0.04} - \overset{''}{1.33}\,T$$
$$\delta\pi = +5.83 + 6.34\,T$$

The definitive values thus become

$$\delta l = +\overset{''}{0.30} - \overset{''}{0.77}\,T$$
$$\delta\pi = +5.59 + 7.31\,T$$

Definitive elements of the four inner planets for the epoch 1850, as inferred from all the data of observation.

88. We have made a fourth solution of the normal equations which give the corrections to the elements of each planet by substituting in those equations the definitive values of all the other quantities, including the values of the secular variations derived from theory. In making this substitution for Mercury, however, the ulterior corrections just found were not applied. The values of the unknowns resulting from this solution are shown in the first column of the next table. From these numbers are derived the definitive elements for 1850, by the following processes:

(α.) By multiplying the unknowns by the appropriate factor given in § 27, we have the corrections of the tabular elements at the mid-epoch of observations for each planet. These corrections are found in the second column.

(β.) The preceding corrections are to be reduced from the respective mid-epochs to 1850. This reduction is found by multiplying the definitive correction to the tabular secular variation by the elapsed interval, and is shown in the third column.

(γ) We next have the value of the tabular elements for the fundamental epoch 1850, January 0, Greenwich mean noon. These numbers are those of LEVERRIER'S tables, with the following modifications:

(δ) The reduction from 1850, January 1, Paris noon, to January 0, Greenwich noon

(ϵ) The corrections to LEVERRIER'S values of the eccentricity and perihelion which are necessary to represent those terms in the perturbations of the mean longitude which depend only upon the sine and cosine of the mean anomaly. The theory is more symmetrical in form when all such terms are included with those of the elliptic motion. In LEVERRIER'S tables they have the following values:

$$\text{Mercury; } \delta v = + 0.030 \sin l - 0.111 \cos l$$
$$\text{Venus; } \quad\quad\quad + 0.010 \quad\quad + 0.037$$
$$\text{Earth; } \quad\quad\quad - 0.067 \quad\quad - 0.098$$
$$\text{Mars; } \quad\quad\quad + 1.061 \quad\quad + 0.718$$

These terms of the longitude may be represented by the following corrections to the elements:

$$\text{Mercury; } \delta e = + 0.058 \quad\quad \delta\pi = 0.0$$
$$\text{Venus; } \quad\quad\quad - 0.012 \quad\quad\quad + 2.3$$
$$\text{Earth; } \quad\quad\quad + 0.054 \quad\quad\quad + 1.4$$
$$\text{Mars; } \quad\quad\quad + 0.613 \quad\quad\quad - 1.0$$

Applying these corrections δ and ϵ to LEVERRIER'S tabular quantities, we have the values of the tabular elements as given in the fourth column. Then applying the preceding corrections we have the definitive values given in the last column.

In some cases this derivation is modified. Instead of using the correction to the perihelion, mean longitude and mean motion of Mercury given by the unknown quantities of the equations, we have used the values for 1850 derived from the discussion of the preceding section.

The quantities which give the position of the node and inclination have been treated in the same way as their secular variations. The symbols J and N indicate values of the unknown quantities related to the corrections of the elements J and N. These unknowns are then changed to corrections of the elements by the factors of §27, and these again to correction of the inclination and node by the equations of §41.

In the case of the node of Venus two values are given. The value (a) is that which follows immediately from the normal equations. If we carry forward the position of the node just derived to the mean epoch of the last two transits of Venus, we find a discrepancy amounting to 2″.04 in the longitude, corresponding to a difference of 0″.121 in the heliocentric latitude. This is considerably larger than the probable error of the results of the observations of the transits. It may, therefore, be questioned whether the latter are not entitled to a greater relative weight than that assigned, owing to the probable systematic errors of the meridian observations. A second value (b) has therefore been derived from the observations of the transits alone. In subsequent investigations we may choose between these two values.

Formation of definitive elements of the four inner planets, for the epoch 1850, January 0, Greenwich mean noon.

Mercury.

	Unknown of equations.	Corr. of element.	Red. to 1850.	Tabular element.	Concluded element.
		″	″	″	″
n	$- .0940$	$- 0.77$	0.0	$538\ 106\ 654.49$	$538\ 106\ 653.72$
e	$- .0741$	$- 0.222$	$- 0.005$	$42\ 409.088$	$42\ 408.861$
π	$+ .6763$	$+ 5.59$	0	$75°\ 7′\ 13.78″$	$75°\ 7′\ 19.37″$
$\imath$	$- .0402$	$+ 0.30$	0	$323\ 11\ 23.53$	$323\ 11\ 23.83$
i	$- .2762\ \mathrm{J} -$	0.64	$- 0.07$	$7\ 0\ 7.71$	$7\ 0\ 7.00$
θ	$- .0001\ \mathrm{N} +$	3.88	$- 0.27$	$46\ 33\ 8.63$	$46\ 33\ 12.24$

Formation of definitive elements, etc.—Continued.

Venus.

Unknown of equations.	Corr. of element.	Red. to 1850.	Tabular element.	Concluded element.
	$''$	$''$	$''$	$''$
n $-$ .1783	$-$ 3.57	0	210 669 165.04	210 669 161.47
e $+$ .1403	$+$ 0.439	$-$ 0.165	1 411.522	1 411.796
			$\overset{\circ}{1}29\;\;\overset{'}{27}\;\;\overset{''}{14.3}$	$\overset{\circ}{1}29\;\;\overset{'}{27}\;\;\overset{''}{34.5}$
π $+$ .0835	$+$ 36.6	-16.4	129 27 14.3	129 27 34.5
ι $-$ .1330	$-$ 0.67	$+$ 0.46	243 57 44.34	243 57 44.13
i $+$ .0968 J	$+$ 0.31	$+$ 0.12	3 23 34.83	3 23 35.26
$\theta\,(a)$ $+$ .0126 N	$-$ 9.39	$+$ 6.63	75 19 52.21	75 19 49.45
$\theta\,(b)$	$-$ 20.36	$+15.56$		75 19 47.41

Earth.

	$''$		$''$	$''$
n	$-$ 1.10		129 602 767.84	129 602 766.74
e	$+$ 0.12		3 459.334	3 459.454
π	$-$ 2.4		$\overset{\circ}{1}00\;\;\overset{'}{2}1\;\;\overset{''}{43.4}$	$\overset{\circ}{1}00\;\;\overset{'}{2}1\;\;\overset{''}{41.0}$
ε	$-$ 0.15		23 27 31.83	23 27 31.68
ι	$+$ 0.02		99 48 18.72	99 48 18.74

Mars.

		$''$		$''$	$''$
n $-$ .1094	$-$ 0.88	0	68 910 105.38	68 910 104.50
e $-$ .1088	$-$ 0.155	$+$ 0.058	19 237.101	19 237.004
π $+$ .1603	$+$ 2.38	$+$ 0.02	$\overset{\circ}{3}33\;\;\overset{'}{17}\;\;\overset{''}{52.47}$	$\overset{\circ}{3}33\;\;\overset{'}{17}\;\;\overset{''}{54.87}$
ι $-$ .4029	$-$ 0.81	$+$ 0.05	83 9 16.92	83 9 16.16
i $-$ .0507 J	$+$ 0.18	$-$ 0.01	1 51 2.28	1 51 2.45
θ $+$ .1135 N	$+$ 6.56	$+$ 1.34	48 23 53.02	48 24 0.92

Definitive values of the secular variations.

89. The definitive values of the secular variations, as inferred from the adopted theories and the concluded values of the masses, are shown in the following table, which gives in detail the parts of which each quantity is made up.

The first four lines of the table give the values of the secular variations as they result from the investigations found in Vol. V, Part IV, of the *Astronomical Papers*, after correcting the mass of each planet by its appropriate factor.

The motion of the perihelion first given, denoted by $D_t\,\pi_1$, is measured along the plane of the orbit itself. The numbers given being divided by the corresponding values of the eccentricity we have the motion of the perihelion itself along the plane. The symbols i_0 and θ_0 represent the inclinations and longitudes of the nodes referred at each epoch to the ecliptic and equinox of 1850, regarded as fixed. The motions of these elements are next to be referred to the fixed ecliptic of the date. So referred, they are designated as $D_t^0 i$ and $D_t^0 \theta$. The transformations to the latter quantities are made by computing an approximate value of the motion of the node due to the motion of the ecliptic alone along the plane of the orbit regarded as fixed.

If we put

i_1, the inclination of the fixed orbit of the planet at any epoch T_0 to the moving ecliptic at any time;

θ_1, the longitude of the corresponding node, Ω_1;

ν, the distance from the node Ω_1 to the instantaneous rotation axis of the orbit at the epoch T_0;

we shall have

$$D_t \nu = \varkappa'' \operatorname{cosec} i_1 \sin (L'' - \theta_1) \qquad (a)$$

If we compute ν_0 and $\varkappa$ from the equations

$$\varkappa \sin \nu_0 = \sin i_0 \, D_t^0 \, \theta_0$$
$$\varkappa \cos \nu_0 = D_t^0 \, i_0$$

and then find $\varDelta \nu$ by integrating the value (a) of $D_t \nu$ from 1850 to the date we shall have

$$\sin i \, D_t^0 \, \theta = \quad \varkappa \sin (\nu_0 + \varDelta \nu)$$
$$D_t^0 \, i = \quad \varkappa \cos (\nu_0 + \varDelta \nu)$$

The change of $D_t \nu$ between 1850 and the extreme epochs has been found so nearly uniform that it was sufficient to multiply its value at the mid-epoch (1675 or 1975) by 2.5 to obtain $\varDelta \nu$.

Next, we have the changes in i and θ due to the motion of the ecliptic, represented by $D_t^1 i$ and $D_t^1 \theta$, and computed by the formula

$$D_t^1 \, i = - \varkappa'' \cos (\nu'' - \theta)$$
$$\sin i \, D_t^1 \, \theta = - \varkappa'' \cos i \sin (\nu'' - \theta)$$

The planetary precession due to the motion of the ecliptic is here omitted, to be afterwards included in the general precession. The sum of the two motions gives the actual variation at each epoch, referred to a fixed equinox.

The motion of θ itself thus found is increased by the general precession, which gives the motion of θ at each epoch.

The motion of the perihelion to be actually used in the tables is equal to the motion of the node from the mean equinox, plus the increase of the arc of the orbit between the node and perihelion. The adopted value of this quantity is found by increasing the motion of π_1 by the following quantities:

1. The change due to the motion of the plane of the orbit.
2. The change due to the motion of the ecliptic.

The formulæ for these two quantities are

$$(1); \; \delta_1 D_t \pi = \qquad \tan \tfrac{1}{2} i \, \sin i \, D_t^0 \, \theta$$
$$(2); \; \delta_2 D_t \pi = \varkappa'' \tan \tfrac{1}{2} i \, \sin (L'' - \theta)$$

3. The excess of motion shown by observations in the case of Mercury and Mars, and computed for all four planets as if they gravitated toward the Sun with a force proportional to r^{-n} where

$$n = 2.000\,000\,16120$$

The values of this correction are

$$
\begin{array}{ll}
\text{Mercury;} & D_t \pi = 43.37 \\
\text{Venus;} & 16.98 \\
\text{Earth;} & 10.45 \\
\text{Mars;} & 5.55 \\
\end{array}
$$

4. The general precession.
5. In the case of the Earth, the motion arising from the action of the Moon, of which the amount is

$$D_t \pi'' = 7''.68$$

But the first two corrections drop out in this case.

The preceding transformations of the secular variations are made with the original values of the elements e and i, as given in *Astronomical Papers*, Vol. V, Part IV, pp. 337, 338.

Secular variations of the elements of the four orbits at the three epochs, 1600, 1850, and 2100, as inferred from the definitively adopted masses.

Mercury.

	1600.	1850.	2100.
$D_t e$	$+$ 4.257	$+$ 4.227	$+$ 4.196
$e\,D_t\,\pi_1$	$+$ 109.524	$+$ 109.498	$+$ 109.475
$D_t^0\,i_0$	$-$ 21.581	$-$ 21.568	$-$ 21.551
$\sin i_0\,D_t^0\,\theta_0$	$-$ 54.891	$-$ 54.969	$-$ 55.049
$D_t^0 i$	$-$ 21.786	$-$ 21.568	$-$ 21.347
$\sin i\,D_t^0\,\theta$	$-$ 54.813	$-$ 54.969	$-$ 55.130
$D_t^1 i$	$+$ 28.884	$+$ 28.333	$+$ 27.785
$\sin i\,D_t^1\,\theta$	$-$ 37.196	$-$ 37.397	$-$ 37.595
$D_t i$	$+$ 7.098	$+$ 6.765	$+$ 6.438
$\sin i\,D_t\,\theta$	$-$ 92.009	$-$ 92.366	$-$ 92.725
$\varDelta\,D_t\,\pi$	$-$ 1.06	$-$ 1.06	$-$ 1.06
$D_t\,\pi$	5593.41	5598.70	5604.02
$D_t\,\theta$	4262.98	4266.12	4269.24

Venus.

	1600.	1850.	2100.
$D_t e$	$-$ 9.959	$-$ 9.866	$-$ 9.772
$e\,D_t\,\pi_1$	$+$ 0.384	$+$ 0.219	$+$ 0.060
$D_t^0 i_0$	$-$ 2.484	$-$ 3.071	$-$ 3.656
$\sin i_0\,D_t^0\,\theta_0$	$-$ 59.005	$-$ 59.112	$-$ 59.229
$D_t^0 i$	$-$ 3.049	$-$ 3.071	$-$ 3.091
$\sin i\,D_t^0\,\theta$	$-$ 58.978	$-$ 59.112	$-$ 59.260
$D_t^1 i$	$+$ 6.690	$+$ 6.695	$+$ 6.697
$\sin i\,D_t^1\,\theta$	$-$ 46.758	$-$ 46.582	$-$ 46.413
$D_t i$	$+$ 3.641	$+$ 3.624	$+$ 3.606
$\sin i\,D_t\,\theta$	$-$ 105.736	$-$ 105.694	$-$ 105.673
$\varDelta\,D_t\,\pi$	$-$ 0.36	$-$ 0.37	$-$ 0.38
$D_t\,\pi$	5090.07	5072.44	5054.92
$D_t\,\theta$	3230.39	3237.98	3245.22

Secular variations of the elements of the four orbits, etc.—Cont'd.

Earth.

	1600.	1850.	2100.
	$''$	$''$	$''$
$D_t\, e''$	$-\quad 8.467$	$-\quad 8.595$	$-\quad 8.727$
$e''\, D_t\, \pi''$	$+\quad 19.293$	$+\quad 19.210$	$+\quad 19.139$
$D_t\, \pi''$	6179.58	6187.41	6195.68
$\varkappa''\, \sin L_0''$	$+\quad 4.370$	$+\quad 5.341$	$+\quad 6.305$
$\varkappa''\, \cos L_0''$	$-\quad 47.113$	$-\quad 46.838$	$-\, .\; 46.550$
$\log \varkappa''$	$.\quad 1.67500$	1.67340	1.67187
L_0''	$174^\circ\, 42'.04$	$173^\circ\, 29'.68$	$172^\circ\, 17'.18$
L''	$171^\circ\, 12'.83$	$173^\circ\, 29'.68$	$175^\circ\, 46'.62$
p_0	5034.91	5036.13	5037.36
p	5018.28	5023.82	5029.38
$D_t\, \varepsilon$	$-\quad 46.761$	$-\quad 46.838$	$-\quad 46.847$

Mars.

	$''$	$''$	$''$
$D_t\, e$	$+\quad 18.775$	$+\quad 18.706$	$+\quad 18.623$
$e\, D_t\, \pi_1$	$+\quad 148.633$	$+\quad 148.707$	$+\quad 148.762$
$D_t^0\, i_0$	$-\quad 28.994$	$-\quad 29.396$	$-\quad 29.803$
$\sin i_0\, D_t^0\, \theta_0$	$-\quad 34.023$	$-\quad 34.012$	$-\quad 34.017$
$D_t^?\, i$	$-\quad 29.482$	$-\quad 29.396$	$-\quad 29.309$
$\sin i\, D_t^0\, \theta$	$-\quad 33.605$	$-\quad 34.012$	$-\quad 34.445$
$D_t^1\, i$	$+\quad 26.964$	$+\quad 27.104$	$+\quad 27.245$
$\sin i\, D_t^1\, \theta$	$-\quad 38.860$	$-\quad 38.551$	$-\quad 38.247$
$D_t\, i$	$-\quad 2.518$	$-\quad 2.292$	$-\quad 2.064$
$\sin i\, D_t\, \theta$	$-\quad 72.465$	$-\quad 72.563$	$-\quad 72.692$
$\varDelta\, D_t\, \pi$	$+\quad 0.08$	$+\quad 0.07$	$+\quad 0.06$
$D_t\, \pi$	6621.51	6623.96	6626.25
$D_t\, \theta$	2776.39	2776.87	2776.63

Secular acceleration of the mean motions.

90. The mean motions of the planets, like that of the Moon, are subject to a secular acceleration arising from the secular variations of the elements of the orbits. The following formulæ for this acceleration are formed by differentiating the known

expressions for the variation of the longitude of the epoch in the theory of the variation of elements. The notation is that of *Astronomical Papers*, Vol. V, Part IV.

We compute for the action of an outer on an inner planet:

$$A = D \, c_3^{(1)}$$

$$B = \frac{1}{8}(D - D^2 - 2\,D^3)\, c_1^{(0)}$$

$$C = \frac{1}{4}(D^2 + D^3)\, c_1^{(0)}$$

$$W = \frac{1}{8}(2 - 9\,D + 3\,D^2 + 4\,D^3)\, c_1^{(1)}$$

Then

$$D_t^2 l_0 = m' \, \alpha \, n \, D_t \{ A \, \sigma^2 + B e^2 - C e'^2 + W e e' \cos(\pi - \pi') \}$$

For the action of an inner on an outer planet we compute

$$A' = -(1 + D)\, c_3^{(1)}$$

$$B' = \frac{1}{4}(D + 2\,D^2 + D^3)\, c_1^{(0)}$$

$$C' = \frac{1}{8}(3\,D + 5\,D^2 + 2\,D^3)\, c_1^{(0)}$$

$$W' = \frac{1}{8}(10 + 3\,D - 9\,D^2 - 4\,D^3)\, c_1^{(1)}$$

$$D_t^2 l_0 = m \, n' \, D_t \{ A' \, \sigma^2 + B' e^2 + C' e'^2 + W' c e' \cos(\pi - \pi') \}$$

The symbol D_t indicates the secular variation of the expression following it produced by the action of all the planets. The unit of time must be the same one in which n is expressed.

The following table gives the results of this computation:

Secular change of the centennial mean motions.

Action of—	Mercury.	Venus.	Earth.	Mars.
	''	''	''	· ''
Venus,	− 0.0426	. . .	− 0.0104	+ 0.0010
Earth,	− 0.0029	+ 0.0128	. . .	+ 0.0119
Mars,	+ 0.0003	− 0.0001	− 0.0012	. . .
Jupiter,	− 0.0039	− 0.0046	− 0.0308	+ 0.0004
Saturn,	− 0.0004	+ 0.0015	+ 0.0021	+ 0.0036
Total,	− 0.0495	+ 0.0096	− 0.0403	+ 0.0169

The measure of time.

91. The fictitious mean Sun whose transit over any meridian defines the moment of mean noon on that meridian is a point on the celestial sphere having a uniform sidereal motion in the plane of the Earth's equator, and a Right Ascension as nearly as may be equal to the Sun's mean longitude. If we put μ for this uniform sidereal motion and add to μ the precession of the equinox in Right Ascension we have for the mean Right Ascension of this fictitious mean Sun

$$\tau = \tau_0 + \mu\,T + 4606''.36\,T + 1''.394\,T^2$$

From §§ 88, 90, and 100 the expression for the Sun's mean longitude, affected by aberration, is found to be

$$L = 279°\,47'\,58''.2 + 129602766''.74\,T + 1''.089\,T^2$$

Equalizing the coefficients of T we find, for the mean Right Ascension of the fictitious mean Sun

$$\tau = 279°\,47'\,58''.2 + 129602766''.74\,T + 1''.394\,T^2$$

This differs from the mean longitude of the actual Sun by the quantity

$$\tau - L = 0''.305\,T^2 = 0^s.020\,T^2$$

This difference is of no importance in the astronomy of our time, but may result in an error of 2^s in the course of one thousand years in the measurement of time by the actual mean sun. We must leave to the astronomers of the future the question how best to meet the question thus arising. Changing to time the expression for τ, the difference or mean excess of sidereal over mean time for the meridian of Greenwich becomes

$$\tau = 18^h\,39^m\,11^s.880 + 24^h\,0^m\,1^s.84449\,t + 0^s.0929\,T^2$$

t being time in Julian years after 1850, January 0, Greenwich mean noon.

Constant of aberration.

92. We first investigate certain fundamental constants connected with the motion of the Sun, Earth, and Moon, on which the precession and nutation depend.

From the adopted value of the solar parallax,

$$\pi = 8''.790,$$

and the adopted velocity of light in kilometers per second,

$$V = 299\ 860,$$

follows for the constant of aberration the value

$$A = 20''.501$$

But if we accept the mean result of the solutions of § 83 as giving the most likely value of the solar parallax, we shall have

$$\pi = 8''.7854$$

Then § 75 will give

$$A = 20''.511$$

as the adjusted value of the constant of aberration.

Mass of the Moon.

93. By means of the equation of § 71 between the lunar inequality P in the motion of the Earth and the mass of the Moon

$$\mu' P = [1.78207]\ \pi$$

we may find a fresh value of the Moon's mass from the values of π and P.

We have found from observation

$$P = 6''.465 \pm .015$$

Thus follows, for the mass of the Moon, when $\pi = 8''.790$,

$$\mu = 1 : 81.32 \pm 0.20$$

Combining this with the value found from the constant of nutation,

$$\mu = 1 : 81.58 \pm 0.20$$

we have, as the definitive mass of the Moon,

$$\mu = 1 : 81.45 \pm 0.15$$

Parallactic inequality of the Moon.

94. From the transformation of HANSEN'S lunar theory in *Astronomical Papers*, Vol. I, it may be concluded that the solar parallax and the parallactic inequality are connected by the relation

$$\text{P. I.} = [1.16242]\frac{1 - \mu}{1 + \mu}\,\pi$$
$$= [1.15176]\,\pi$$

Hence we have, for the coefficient of the parallactic inequality of the Moon, corresponding to $\pi = 8''.790$,

$$124''.66$$

Here the inequality is that in ecliptic longitude.

The centimeter-second system of units.

95. There are certain methods in physics by which the next step in the course of our researches will be guided. The adoption of a system of absolute units has simplified the methods and conceptions of physics to such an extent that we may find it advantageous to introduce a similar system into those investigations of astronomy which are closely connected with that science.

The fundamental units most widely adopted are the centimeter as the unit of length, the gram as the unit of mass, and the second as the unit of time. There is, however, an insuperable difficulty in the way of introducing the gram, or any other arbitrary terrestrial unit of mass, into astronomy, from the fact that the astronomical masses with which we are concerned can not be determined with sufficient precision in units of terrestrial mass. It is, therefore, quite common in celestial mechanics to regard the unit of mass as arbitrary, and to multiply this arbitrary unit by a factor which will represent its attractive force upon a unit particle at unit distance. The introduction of this factor is, however, needless. It is simpler to adopt the course of DELAUNAY and many other writers, and regard the unit of mass as a derived one, based on the units of time and length, by defining it as that mass which will attract an equal mass at unit distance with force

unity. In this definition the unit of force retains its physical
meaning, as that force which, acting on unit mass, will produce a unit of acceleration in a unit of time.

The number of fundamental units is then reduced to two,
those of time and length, and the unit mass becomes a derived
one of dimensions,

$$M = L^3 T^2$$

The centimeter as a unit of length would be inconveniently
small for astronomical purposes, if we had to deal mainly with
natural numbers, but it causes no inconvenience in logarith-
mic computations, and has the advantage of being assimilated
directly to the centimeter-gram-second system in physics.
We shall therefore adopt it, expressing our results, however,
in terms of other units whenever convenience will thereby be
gained.

I shall make clear this assimilation and the use of the unit
of mass as a derived one, by calling this the centimeter-
second system.

In the latter the definitions of units in the centimeter-
gram-second system will remain unchanged, except that the
derived unit of mass must be substituted for the gram. The
dimensions of units in the centimeter-second system will be
found by making the above substitution for M in the expressions for those of the centimeter-gram-second system.

Masses of the Earth and Moon in centimeter-second units.

96. A fundamental quantity in the centimeter-second system
is the mass of the Earth. This mass will be by definition the
force of gravity of the Earth, if concentrated in a point at the
distance of one centimeter. Were the Earth a sphere of known
dimensions, it could be readily determined through the force
of gravity at any point on its surface. This being not the case,
we shall proceed on the accepted approximate theory that the
geoid is an ellipsoid of revolution, and that the force of gravity
at a point the sine of whose latitude is 1 : $\sqrt{3}$, is the same as
if the mass of the Earth were concentrated in its center.

The determination of this constant with astronomical precision is a difficult and we might say hitherto an insoluble prob-

lem, owing to the heterogeneity of the Earth and the absence
of determinations of the force of gravity over the surface of the
ocean. Although the limits of uncertainty thus arising can
not be set with any approach to precision, I do not think they
are such as to greatly impair the astronomical results which
are to be derived from them. Investigations in geodesy not
being practicable in the present work, I have, mainly from a
study of the work of G. W. HILL,[*] assumed for the length of
the seconds pendulum at the point the sine of whose latitude
is $1 : \sqrt{3}$, which I shall call the mean latitude,

$$L_1 = 99.2715$$

With this we may compare HELMERT'S expression for the
length of the seconds pendulum in terms of the latitude

$$L = 0^m.990918 (1 + .005310 \sin^2\varphi)$$

which gives

$$L_1 = 99.2688$$

From these values of L_1 we have:

	HILL.	HELMERT.
Gravity at mean latitude,	979.770	979.745
Correction for centrifugal force,	2.260	2.260
Attraction of the Earth,	982.030	982.005

I also accept as the result of CLARKE'S investigation of 1880,

Equatorial radius of the Earth,	6378249^m
Reduction to mean latitude,	7245
Mean radius of the Earth,	6371004

From HILL'S and HELMERT'S numbers follows:

Logarithm mass of Earth expressed in centimeter-second units.

HILL.	HELMERT.
20.600541.	20.600530.

---

[*] *Astronomical Papers*, Vol. III, p. 339.

From the adopted ratio of the mass of the Moon to that of the Earth:

$$\mu = 1 : 81.45$$

follows

Logarithm of the mass of the Moon in centimeter-second units,

18.68965.

Parallax of the Moon.

97. From these results the distance of the Moon and the relation between the mass and distance of the sun follow in a very simple way. By the formulæ of elliptic motion it follows that when we put

m, m′, the masses of any two bodies revolving around each other in virtue of their mutual gravitation;

a, the semimajor axis of the relative orbit, which would be the actual distance if the motion were circular;

n, their mean angular motion in unit of time;

we have the relation

$$a^3 n^2 = m + m'$$

This relation is rigorous and independent of the adopted units of length and time, provided we define the unit of mass in the way already done. It follows that if the Moon in its revolution around the Earth were not subject to disturbance, its mean motion in one second, and its distance expressed in centimeters, would be connected by the relation

$$\text{Log } a^3 n^2 = \log m'' (1 + \mu) = 20.605841$$

In the theories of DELAUNAY and ADAMS the quantity *a*, as determined by this equation, is accepted as a fundamental element, and it is shown that in consequence of the perturbations produced by the Sun the constant Π_0 of the Moon's horizontal parallax is connected with *a* by the relation

$$a \sin \Pi_0 = 1.000907 \, \rho$$

ρ being the radius of the Earth corresponding to Π_0

From the mean sidereal motion of the Moon in a Julian century

$$1336.85136 \text{ revolutions}$$

we find, for the co-logarithm of the motion in arc in one second

$$\log \frac{1}{n} = 5.574841$$

and thus have for the undisturbed mean distance of the Moon in centimeters

$$\log a = 10.585174$$

and hence

$$\log \sin \Pi_0 = 8.219921$$

$$\begin{array}{lr} & {}' \quad {}'' \\ \Pi_0 = & 57\ 2.68 \\ \text{Red. to sine,} & -\ .16 \\ \text{Constant of sin } \pi \text{ in arc,} & 57\ 2.52 \end{array}$$

Using HELMERT'S length of the seconds pendulum we should have found for this constant

$$3422''.55$$

Mass and parallax of the Sun.

98. In the case of the motion of the center of gravity of the Earth and Moon around the Sun the relation of §97 becomes

$$a'^3 n'^2 = M_1 + m'' (1 + \mu)$$

M_1 being the mass of the Sun. Replacing a' by π, the parallax of the Sun, and ρ the radius of the Earth, we find for the ratio M of the mass of the Sun to the sum of the masses of the Earth and Moon

$$M = \frac{\rho^3 n'^2}{m'' (1 + \mu) \sin^3 \pi} - 1$$

$$\log M \pi^3 = 8.349674$$

The values of M corresponding to certain values of the mean equatorial horizontal parallax of the Sun are as follows:

Π	M
8.780	330514
8.785	329951
8.790	329388
8.795	328827
8.800	328266

Nutation and mechanical ellipticity of the Earth.

99. Regarding the mass of the Moon as known, we now utilize the equations of § 67 to obtain the constant of nutation and the mechanical ellipticity of the Earth. The last two of these equations give, for the absolute precessional constant, when the Julian year is the unit of time,

$$ \mathrm{P} = \left[[5.975052]\,_1\frac{\mu}{1 + \mu} + 5310''.0 \right] \frac{\mathrm{C} - \mathrm{A}}{\mathrm{C}} $$

We have found, in § 66, for a Julian year

$$ \mathrm{P} = 54''.8990 $$

We then have, for the mechanical ellipticity of the Earth,

$$ \frac{\mathrm{C} - \mathrm{A}}{\mathrm{C}} = 0.0032753 $$

We also have, from the first equation of § 66, for the constant of nutation for 1850

$$ \mathrm{N} = 9''.214 $$

For the parts of the precessional constant which arise from the action of the Sun and of the Moon, respectively, we have—

	''
Action of the Sun	17.3919
Action of the Moon	37.5071

Precession.

100. In order to develop the terms of the precession and obliquity to higher powers of the time, I have extended their computation one step backward and forward from the three fundamental epochs, by extrapolation of $\varkappa$ and L. The results are as follows:

Motion of the ecliptic and equator.

Year.	log. $\varkappa$	L	$D_t \varepsilon$	n
		° '	''	''
1350	1.67666	168 56.13	− 46.613	2009.05
1600	1.67500	171 12.84	− 46.761	2006.92
1850	1.67340	173 29.68	− 46.838	2004.79
2100	1.67187	175 46.63	− 46.847	2002.66
2350	1.67039	178 3.50	− 46.789	2000.52

Centennial precessions for tropical centuries.

Year.	In longitude — Lunisolar.	Planetary.	General.	In Right Ascension.
	''	''	''	''
1350	5033.58	− 20.94	5012.64	4592.41
1600	5034.80	− 16.63	5018.17	4599.38
1850	5036.02	− 12.31	5023.71	4606.36
2100	5037.25	− 7.98	5029.27	4613.35
2350	5038.49	− 3.67	5034.82	4620.32

From these values we have the following general expressions:

	'' ''
Annual precession in Right Ascension;	$46.0636 + 0.0279\,T$
Annual precession in longitude;	$50.2371 + 0.0222\,T$
Centennial precession in longitude;	$5023.71 + 2.218\,T$
Total precession from 1850;	$5023.71\,T + 1.109\,T^2$

Mean obliquity of the ecliptic.

101. The expression for the mean obliquity when T is counted from 1900 is —

$$\varepsilon = 23° \; 27' \; 8''.26 - 46''.845\,T - 0''.0059\,T^2 + 0''.00181\,T^3$$

Tables of the mean obliquity at different epochs.

Year.	Obliquity.	Year.	Obliquity.
	° ′ ″		° ′ ″
1600	23 29 28.69	− 2500	23 58 44.00
1650	29 5.31	− 2000	55 38.99
1700	28 41.91	− 1500	52 23.10
1750	28 18.51	− 1000	48 57.70
1800	27 55.10	− 500	45 24.14
1850	27 31.68	0	41 43.78
1900	27 8.26	+ 500	37 57.97
1950	26 44.84	1000	34 8.07
2000	26 21.41	1500	30 15.43
2050	25 57.98	2000	26 21.41
2100	23 25 34.56	2500	23 22 27.37

Relative positions of the equator and ecliptic at different dates.

102. The motions expressed in the preceding tables are, for the most part, purely instantaneous ones, referred to the planes of the ecliptic and equator of each separate epoch. For the reduction of the places of the fixed stars from one epoch to another, it is necessary to know the relative position of the planes of the equator or ecliptic at the two epochs. We shall therefore derive the fundamental quantities which express the position of the equator and the ecliptic at any one epoch relatively to their positions at a fundamental epoch taken at pleasure. The latter we shall call zero-position. Then, the zero equator and ecliptic are those of the fundamental epoch; the equator and ecliptic simply those of any other varying epoch. So far as convenient, and as conducive to ease in comparing our results with former ones, we shall use the notation of BESSEL.

To derive the equations for the motions, let us consider the following four points of the celestial sphere:

E_0, the pole of the zero ecliptic.

E, the pole of the actual ecliptic.

P_0, the pole of the zero equator.

P, the pole of the actual equator.

We put,

$\varepsilon_1 = P E_0$, the obliquity of the equator to the zero ecliptic;

$k = E E_0$, the inclination of the two ecliptics;

Π_0, the longitude of the node of the ecliptic on the zero ecliptic, measured from the zero equinox of the date;

Π_1, the longitude of the same node, measured from the actual equinox;

λ, the arc of the equator intercepted between the two ecliptics, or the planetary precession on the equator;

ψ, the total lunisolar precession on the zero ecliptic from the zero epoch to the actual epoch;

n, the rate of motion of the pole of the equator;

τ, the time, expressed in units of 250 years from the zero epoch to any other epoch.

The position of the variable point E is defined by the quantities k and Π_0 or Π_1, which are themselves to be determined through the values of $\varkappa$ and L of § 100.

The position of the variable point P is determined by the condition that its motion is constantly at right angles to the arc EP, and its velocity measured on the arc of a great circle is given by the equation

$$\frac{d s}{d t} = n = P \sin \varepsilon \cos \varepsilon \qquad (a)$$

The positions of the equator and equinox relative to the zero equator and ecliptic are then determined by the quantities ε_1, ψ and λ. The spherical triangle P E_0 E gives the following equations:

$$\frac{\sin \lambda}{\sin k} = \frac{\sin \Pi_1}{\sin \varepsilon_1} = \frac{\sin \Pi_0}{\sin \varepsilon}$$

During a period of several centuries the quantities k and λ are so small that no distinction is necessary between them and their sines. We may therefore put

$$\lambda = k \sin \Pi_1 \operatorname{cosec} \varepsilon_1 = k \sin \Pi_0 \operatorname{cosec} \varepsilon \qquad (b)$$

We also have, from the law of motion of the pole of the equator,

$$D_t \varepsilon_1 = n \sin \lambda$$
$$D_t \psi = n \cos \lambda \operatorname{cosec} \varepsilon_1 \qquad (c)$$

As the value of ε_1 does not change by $0''.6$ from one epoch to another, we may, without appreciable error, use ε_0 for ε_1 in the formulæ (*b*) and (*c*). To use these equations, we first obtain *k* and $\varPi_1$ from the secular motion of the ecliptic, while *n* is computed for any epoch from the formula (*a*). We then easily develop the values of ε_1 and ψ in powers of the time by the equations (*c*). The values of *n* have no reference to any special coordinates. From the table of § 100 it will be seen that we may put

$$n = 2004''.79 - 2''.13 \, \tau'$$

τ' being counted from 1850.

To find the value of $\varPi_1$ in each case, we remark that the instantaneous values of L given in § 100 show that the instantaneous node, or intersections of two consecutive ecliptics, moves with so near an approach to uniformity that we may take for the actual node between the ecliptics of any two epochs τ_1 and τ_2 the mean of the instantaneous nodes for those two epochs. For example, let it be required to find the value of $\varPi_1$ for the node of the ecliptic of 2100 on that of 1850. We have

For 2100 . . . . L $= 175°\ 46'.63$
For 1850, referred to eq. of 2100 . L $= 176°\ 59'.13$
Concluded value of $\varPi_1$ . . . . $\varPi_1 = 176°\ 22'.9$

As the basis of our work we have computed the required quantities for the zero ecliptics of 1600, 1850, and 2100, respectively. The values of *k* and $\varPi_1$ for the ecliptics of two hundred and fifty years before and after these epochs are as follows:

Zero epoch.	− 250 Y		+ 250 Y	
	k	$\varPi_1$	*k*	$\varPi_1$
	$''$	$°\quad '$	$''$	$°\quad '$
1600	−118.48	168 20.0	+118.07	174 5.9
1850	−118.07	170 36.7	+117.64	176 22.9
2100	−117.64	172 53.4	+117.23	178 39.9

Changing the unit of time to two hundred and fifty years, the equations (a) (b) and (c) give the following values of the derivatives of ε_1 and ψ:

Zero-epoch.	$D_\tau\varepsilon$		$D_\tau\psi$	
	-250 Y	$+250$ Y	-250 Y	$+250$ Y
	$''$	$''$	$''$	$''$
1600	-1.4636	$+0.7400$	12600.33	12573.65
1850	-1.1768	$+0.4527$	12603.44	12576.65
2100	-0.8898	$+0.1665$	12606.57	12579.71

At the respective epochs $D_\tau\varepsilon_1$ vanishes, and $D_\tau\psi$ has the values of the lunisolar precession in longitude (§ 100).

Developing in powers of τ we have the following results:

Zero-epoch. ° ′ ″ ″ ″
$$1600; \quad \varepsilon_1 = 23\ 29\ 28.69 + 0.5509\ \tau^2 - 0.1206\ \tau^3$$
$$1850; \quad \varepsilon_1 = 23\ 27\ 31.68 + 0.4074 \qquad\ - 0.1207$$
$$2100; \quad \varepsilon_1 = 23\ 25\ 34.56 + 0.2641 \qquad\ - 0.1206$$

$$\qquad\qquad\qquad\qquad\quad '' \qquad\quad ''$$
$$1600; \quad \psi = 12587.00\ \tau - 6.67\ \tau^2$$
$$1850; \quad \psi = 12590.05 \quad - 6.70$$
$$2100; \quad \psi = 12593.14 \quad - 6.72$$

$$\qquad\qquad\qquad '' \qquad\quad ''$$
$$1600; \quad \lambda = 45.28\ \tau - 14.83\ \tau^2$$
$$1850; \quad \lambda = 33.52 \quad - 14.86$$
$$2100; \quad \lambda = 21.75 \quad - 14.88$$

These values of ε_1 and ψ completely fix the position of the equator at the time τ relative to the zero ecliptic and equinox. For the reduction of coordinates from one epoch to another we must express the position of the equator at the time τ. We consider the triangle $P\,E_0\,P_0$, of which the sides and opposite angles are designated

Sides, $\qquad\qquad\qquad \varepsilon_0 \qquad\quad \varepsilon_1 \qquad \theta$
Opposite angles, $\quad 90° - \zeta \quad 90° - \zeta_1 \quad \psi$

If, in the Gaussian relations between the parts of this triangle, we put

$$\sin \tfrac{1}{2}\,(\varepsilon_1 - \varepsilon_0) = \tfrac{1}{2}\,(\varepsilon_1 - \varepsilon_0) = \tfrac{1}{2}\,\Delta\varepsilon$$

and regard the cosine of this angle as unity, we have

$$\tan \tfrac{1}{2}(\zeta + \zeta_1) = \cos \tfrac{1}{2}(\epsilon_1 + \epsilon_0)\tan \tfrac{1}{2}\psi$$

$$\tan \tfrac{1}{2}(\zeta - \zeta_1) = \frac{\Delta\epsilon}{2\sin \tfrac{1}{2}(\epsilon_1 + \epsilon_0)\tan \tfrac{1}{2}\psi}$$

If we develop the differences between the tangent and the arc we find from these equations

$$\zeta + \zeta_1 = \psi \cos \tfrac{1}{2}(\epsilon_1 + \epsilon_0)(1 + \tfrac{1}{12}\psi^2 \sin^2 \epsilon_0)$$

$$\zeta - \zeta_1 = \frac{2\,\Delta\epsilon}{\psi \sin \tfrac{1}{2}(\epsilon_1 + \epsilon_0)}(1 - \tfrac{1}{12}(z_0^2 + \psi^2))$$

where we put z_0 for the approximate value of $\zeta - \zeta_1$

For the inclination θ of the mean equator of the epoch τ to the zero equator, we have the equation

$$\sin \theta = \frac{\sin \epsilon_0 \sin \psi}{\cos \zeta}$$

and then, by developing in powers of θ and ψ, we find

$$\theta = \frac{\psi \sin \epsilon_0}{\cos \zeta}(1 - \tfrac{1}{6}\psi^2 \cos^2 \epsilon_0)$$

$$= \psi \sin \epsilon_0 (1 + \tfrac{1}{2}\zeta^2)(1 - \tfrac{1}{6}\psi^2 \cos^2 \epsilon_0)$$

We thus find

Zero-epoch.		″	″	″
1600;	$\zeta + \zeta_1 =$	11543.79 τ	$- 6.12\,\tau^2$	$+ 0.57\,\tau^3$
1850;		11549.44	$- 6.14$	$+ 0.57$
2100;		11555.12	$- 6.16$	$+ 0.58$

		″	″	
1600;	$\zeta - \zeta_1 =$	45.29 τ	$- 9.92\,\tau^2$	
1850;		33.53	$- 9.93$	
2100;		21.76	$- 9.94$	

		″	″	″
1600;	$\theta =$	5017.30 τ	$- 2.66\,\tau^2$	$- 0.64\,\tau^3$
1850;		5011.97	$- 2.67$	$- 0.64$
2100;		5006.64	$- 2.67$	$- 0.65$

To show the significance of the preceding quantities, consider once more the spherical quadrangle $P_0\,E_0\,EP$. Let these letters represent the positions of the poles on the celestial sphere at any two epochs. In this quadrangle we shall have

$$\text{Angle } E_0 \, P_0 \, E \; = 90^\circ - \zeta_1$$
$$\text{Angle } E \; \; P \; \; P_0 = 90^\circ - \zeta + \lambda$$
$$\text{Side } P_0 \, P = \theta$$

Let S be the position of a star on the celestial sphere. Its polar distances at the two epochs will be $P_0 S$ and $P S$ and its Right Ascensions will be determined by the angles P_0 and P of the triangle $S P_0 P$.

Thus, if the Right Ascension and Declination of S are given for one epoch, we can find it for the other epoch by the solution of the triangle $S P P_0$ when we have given the values of the quantities θ, ζ_1, and $\zeta + \lambda$.

To find the values of these quantities from the preceding formula, let T be the zero-epoch, expressed in calendar years, and let τ be the interval between the two epochs, taken positively when the zero-epoch is the earlier one, and negatively when it is the later one. We interpolate the coefficients of τ and its powers from the preceding formula to the epoch T. Then by substituting the value of τ in the formula we shall have the values of the required quantities, and hence the data for reducing the position of S from one epoch to the other.